Stochastic Models of Financial Mathematics

Optimization in Insurance and Finance Set

coordinated by
Nikolaos Limnios and Yuliya Mishura

Stochastic Models of Financial Mathematics

Vigirdas Mackevičius

ELSEVIER

First published 2016 in Great Britain and the United States by ISTE Press Ltd and Elsevier Ltd

ISTE Press Ltd
27-37 St George's Road
London SW19 4EU
UK

www.iste.co.uk

Elsevier Ltd
The Boulevard, Langford Lane
Kidlington, Oxford, OX5 1GB
UK

www.elsevier.com

Notices

Knowledge and best practice in this field are constantly changing. As new research and experience broaden our understanding, changes in research methods, professional practices, or medical treatment may become necessary.

Practitioners and researchers must always rely on their own experience and knowledge in evaluating and using any information, methods, compounds, or experiments described herein. In using such information or methods they should be mindful of their own safety and the safety of others, including parties for whom they have a professional responsibility.

To the fullest extent of the law, neither the Publisher nor the authors, contributors, or editors, assume any liability for any injury and/or damage to persons or property as a matter of products liability, negligence or otherwise, or from any use or operation of any methods, products, instructions, or ideas contained in the material herein.

For information on all our publications visit our website at http://store.elsevier.com/

British Library Cataloguing-in-Publication Data
A CIP record for this book is available from the British Library
Library of Congress Cataloging in Publication Data
A catalog record for this book is available from the Library of Congress
ISBN 978-1-78548-198-7

Printed and bound in the UK and US

Contents

Preface

These lecture notes are based on a graduate course given for several years at Vilnius University as part of the Master's program "Financial and Actuarial Mathematics". They are intended to give a short introduction to continuous-time financial models including Black–Scholes and interest rate models. We assume the reader to be familiar with the basics of probability theory in the scope of a standard elementary course. Some basic knowledge of stochastic integration and differential equations theory is preferable, although, formally, all the preliminary information is given in part 1 of the lecture notes.

Though a large number of books and textbooks have influenced the writing of these notes, the short reference list includes only the literature directly used by the author.

The author would like to thank a large number of master's students of the Faculty of Mathematics and Informatics of Vilnius University. Thanks to them, the book contains significantly fewer mistakes.

Vigirdas MACKEVIČIUS
Vilnius, September 2016

Notations

$\mathbb{N}$	The set of positive integers $\{1, 2, \ldots\}$
$\overline{\mathbb{N}}$	$\mathbb{N} \cup \{+\infty\}$
$\mathbb{N}_+$	$\mathbb{N} \cup \{0\}$
$\mathbb{R}$	Real line $(-\infty, +\infty)$
$\overline{\mathbb{R}}$	Extended real line $\mathbb{R} \cup \{-\infty, +\infty\}$
$\mathbb{R}_+$	The set of non-negative real numbers $[0, +\infty)$
$x \vee y$	$\max\{x, y\}$
$x \wedge y$	$\min\{x, y\}$
$\mathbf{1}_A$	The indicator of a set (an event) A: $\mathbf{1}_A(x) = 1$ for $x \in A$; $\mathbf{1}_A(x) = 0$ for $x \in A^c$
$\mathbf{E}(X)$, $\mathbf{E}\,X$	The expectation (or mean) of a random variable X
$\mathrm{Var}(X)$, $\mathrm{Var}\,X$	The variance of a random variable X
$N(a, \sigma^2)$	Normal distribution with expectation a and variance σ^2
$X \sim N(a, \sigma^2)$	A random variable X with distribution $N(a, \sigma^2)$
φ	The standard normal density (the probability density of $X \in N(0, 1)$): $\varphi(x) = \frac{1}{\sqrt{2\pi}} e^{-x^2/2}$, $x \in \mathbb{R}$
$\mathcal{N}$	The standard normal distribution function: $\mathcal{N}(x) = \int_{-\infty}^{x} \varphi(y)\,\mathrm{d}y$, $x \in \mathbb{R}$
$X \overset{d}{=} Y$	Random variables X and Y are identically distributed
$X \perp\!\!\!\perp Y$	Random variables X and Y are independent

$\mathcal{H}_t$ The history (or the past) of Brownian motion B up to moment t (Definition 1.2)

$H^2[0,T]$ The set of adapted processes $X = \{X_t,\, t \in [0,T]\}$ with $\|X\|_{H^2} = (\mathbf{E}\int_0^T X_s^2 \mathrm{d}s)^{1/2} < +\infty$ (Definition 1.3)

$\widehat{H}^2[0,T]$ The set of adapted processes $X = \{X_t,\, t \in [0,T]\}$ with $\mathbf{P}\{\int_0^T X_s^2 \mathrm{d}s < +\infty\} = 1$ (Definition 1.6)

$\int_0^t Y_s \mathrm{d}X_s$ The stochastic (Itô) integral of a process Y with respect to an Itô process X (Definitions 1.5 and 1.11)

$\langle X, Y \rangle$ The (quadratic) covariation of Itô processes X and Y (Definition 1.12)

$\langle X \rangle = \langle X, X \rangle$ The quadratic variation of an Itô process X (Definition 1.12)

Overview of the Basics of Stochastic Analysis

1.1. Brownian motion

We suppose that all random variables and processes considered are defined on one probability space $(\Omega, \mathcal{F}, \mathbf{P})$.

DEFINITION 1.1.– *Brownian motion* is a continuous random process with the following properties:

1) $B_0 = 0$, and $B_t - B_s \sim N(0, t - s), 0 \leqslant s < t$;

2) for any $0 \leqslant t_0 < t_1 < t_2 < \cdots < t_k < \infty$, the random variables $B_{t_1} - B_{t_0}, B_{t_2} - B_{t_1}, B_{t_3} - B_{t_2}, \ldots, B_{t_k} - B_{t_{k-1}}$ are independent.

THEOREM 1.1.– With probability one, the trajectories of Brownian motion are non-differentiable at all points $t \geqslant 0$.

THEOREM 1.2.– Let $M_t := \sup_{s \leqslant t} B_s$.[1] Then, the joint distribution function of (B_t, M_t) is

$$\mathbf{P}\{B_t \leqslant x, M_t \leqslant y\} = \begin{cases} \mathcal{N}\left(\frac{x}{\sqrt{t}}\right) - \mathcal{N}\left(\frac{x-2y}{\sqrt{t}}\right), & y \geqslant 0, \ x \leqslant y, \\ \mathcal{N}\left(\frac{y}{\sqrt{t}}\right) - \mathcal{N}\left(\frac{-y}{\sqrt{t}}\right), & y \geqslant 0, \ x \geqslant y, \\ 0, & y < 0, \end{cases}$$

where $\mathcal{N}$ is the distribution function of the standard normal variable.

---

1 Maximum process of Brownian motion.

In particular, $M_t \overset{d}{=} |B_t|$ for all $t \geqslant 0$.

PROOF.– We shall use the so-called *symmetry principle* of Brownian motion. A random variable X is called symmetric if $X \overset{d}{=} -X$. The increments in Brownian motion $B_t - B_\tau$, $t > \tau$, are distributed by the normal law $N(0, t - \tau)$ and therefore are symmetric. The symmetry principle of Brownian motion says that these increments also remain symmetric when τ is not a fixed time moment, but rather the first (random) time of hitting some level $y \in \mathbb{R}$, i.e. when

$$\tau = \tau_y = \min\{s \geqslant 0 : B_s = y\}.$$

(Intuitively, this fact, though rather obvious, is not trivial and is a special case of the more general Markov property of Brownian motion.) More precisely, the random variables

$$\left(B_t - B_{\tau_y}\right)\mathbf{1}_{\{\tau_y \leqslant t\}} = \begin{cases} B_t - y, & \tau_y \leqslant t, \\ 0, & \tau_y > t, \end{cases}$$

are symmetric. Now, let us consider separate cases.

Case 1. $y \geqslant 0$, $x \leqslant y$. In this case, since $B_{\tau_y} = y$ on the event $\{\tau_y \leqslant t\} = \{M_t \geqslant y\}$, we have

$$\begin{aligned}
\mathbf{P}\{B_t \leqslant x, M_t \geqslant y\} &= \mathbf{P}\{B_t \leqslant x, \tau_y \leqslant t\} \\
&= \mathbf{P}\{B_t - B_{\tau_y} \leqslant x - y, \tau_y \leqslant t\} \\
&= \mathbf{P}\{B_t - B_{\tau_y} \geqslant -(x - y), \tau_y \leqslant t\} \\
&= \mathbf{P}\{B_t - y \geqslant -(x - y), \tau_y \leqslant t\} \\
&= \mathbf{P}\{B_t \geqslant 2y - x, \tau_y \leqslant t\} = \mathbf{P}\{B_t \geqslant 2y - x\}, \qquad [*]
\end{aligned}$$

where the last equality follows from $2y - x \geqslant y$, which implies that $\{B_t \geqslant 2y - x\} \subset \{B_t \geqslant y\} \subset \{\tau_y \leqslant t\}$. From $[*]$, we have

$$\begin{aligned}
\mathbf{P}\{B_t \leqslant x, M_t \leqslant y\} &= \mathbf{P}\{B_t \leqslant x\} - \mathbf{P}\{B_t \leqslant x, M_t \geqslant y\} \\
&= \mathbf{P}\{B_t \leqslant x\} - \mathbf{P}\{B_t \geqslant 2y - x\} \\
&= \mathbf{P}\{B_t \leqslant x\} - \mathbf{P}\{B_t \leqslant x - 2y\} \qquad [B_t \text{ is a symmetric r.v.}] \\
&= \mathcal{N}\left(\frac{x}{\sqrt{t}}\right) - \mathcal{N}\left(\frac{x - 2y}{\sqrt{t}}\right). \qquad [**]
\end{aligned}$$

Case 2. $0 \leqslant y \leqslant x$. As $M_t \geqslant B_t$, in this case, we have

$$\mathbf{P}\{B_t \leqslant x,\ M_t \leqslant y\} = \mathbf{P}\underbrace{\{y < B_t \leqslant x,\ M_t \leqslant y\}}_{\varnothing} + \mathbf{P}\{B_t \leqslant y,\ M_t \leqslant y\}$$

$$= \mathbf{P}\{B_t \leqslant y,\ M_t \leqslant y\}.$$

We find the last probability by taking $x = y$ in equation $[**]$.

Case 3. $y < 0$. In this case, $\mathbf{P}\{B_t \leqslant x,\ M_t \leqslant y\} = 0$ since $M_t \geqslant 0$.

In particular, for all $x \geqslant 0$, using equation $[*]$ with $y = x$ and the obvious fact that $\{B_t \geqslant x\} \subset \{M_t \geqslant x\}$, we get

$$\mathbf{P}\{M_t \geqslant x\} = \mathbf{P}\{B_t \leqslant x,\ M_t \geqslant x\} + \mathbf{P}\{B_t \geqslant x,\ M_t \geqslant x\}$$

$$= \mathbf{P}\{B_t \geqslant x\} + \mathbf{P}\{B_t \geqslant x\}$$

$$= \mathbf{P}\{B_t \geqslant x\} + \mathbf{P}\{B_t \leqslant -x\} = \mathbf{P}\{|B_t| \geqslant x\}. \qquad \square$$

COROLLARY 1.1.–

– The joint distribution density of the vector (B_t, M_t) is

$$p_t(x, y) = \frac{2(2y - x)}{\sqrt{2\pi t^3}} e^{-(2y-x)^2/2t}, \quad y \geqslant x^+.$$

– The distribution density of the maximum process M_t is

$$p_t(y) = \frac{2}{\sqrt{2\pi t}} e^{-y^2/2t}, \quad y \geqslant 0.$$

– The distribution density of the hitting time $\tau_y = \min\{s \geqslant 0 : B_s = y\}$ $(y > 0)$ is

$$p_y(t)] = \frac{y}{\sqrt{2\pi t^{3/2}}} e^{-y^2/(2t)}, \quad t \geqslant 0.$$

PROOF.– As $M_t \geqslant \max\{B_t, 0\}$, the joint density of (B_t, M_t) is the mixed second-order derivative of the joint distribution function found in theorem 1.2 in the domain $y \geqslant x^+$ (elsewhere, the density equals zero). Denoting by $\varphi(x) = \frac{1}{\sqrt{2\pi}}e^{-x^2/2}$ the standard normal density, we have

$$p_t(x,y) = \frac{\partial^2}{\partial x \partial y}\left(\mathcal{N}\left(\tfrac{x}{\sqrt{t}}\right) - \mathcal{N}\left(\tfrac{x-2y}{\sqrt{t}}\right)\right)$$

$$= \frac{\partial}{\partial x}\left(\frac{2}{\sqrt{t}}\varphi\left(\tfrac{x-2y}{\sqrt{t}}\right)\right)$$

$$= \frac{2}{t}\varphi'\left(\tfrac{x-2y}{\sqrt{t}}\right) = \frac{2(2y-x)}{t^{3/2}}\varphi\left(\tfrac{x-2y}{\sqrt{t}}\right) \qquad \left[\varphi'(x) = -x\varphi(x)\right]$$

$$= \frac{2(2y-x)}{\sqrt{2\pi t^3}}e^{-(2y-x)^2/2t}, \quad y \geqslant x^+.$$

The densities of M_t and τ_y can be found by differentiating the equality

$$\mathbf{P}\{M_t \geqslant y\} = \mathbf{P}\{\tau_y \leqslant t\} = \mathbf{P}\{|B_t| \geqslant y\} = 2\left(1 - \mathcal{N}\left(\tfrac{y}{\sqrt{t}}\right)\right)$$

with respect to y and t, respectively.   □

Denote[2]

$$m_t := \inf_{s \leqslant t} B_s.$$

Similarly or by using the equality $m_t = -\sup_{s \leqslant t}(-B_s)$, we obtain that, for all $t \geqslant 0$, $m_t \stackrel{d}{=} -|B_t|$ and that the joint distribution density of (B_t, m_t) is

$$\mathbf{P}\{B_t \geqslant x, m_t \geqslant y\} = \begin{cases} \mathcal{N}\left(\tfrac{-x}{\sqrt{t}}\right) - \mathcal{N}\left(\tfrac{2y-x}{\sqrt{t}}\right), & y \leqslant 0, x \geqslant y, \\ \mathcal{N}\left(\tfrac{-y}{\sqrt{t}}\right) - \mathcal{N}\left(\tfrac{y}{\sqrt{t}}\right), & y \leqslant 0, x \leqslant y, \\ 0, & y > 0. \end{cases}$$

[1.1]

As before, by iterate differentiation of [1.1] we obtain the joint density of (B_t, m_t):

$$p_t(x,y) = \begin{cases} \frac{2(x-2y)}{\sqrt{2\pi t^3}}e^{-(x-2y)^2/2t}, & y \leqslant \min\{x,0\} = -x^-, \\ 0 & \text{otherwise.} \end{cases}$$

[1.2]

---

2 The minimum process of Brownian motion.

THEOREM 1.3.– Let B_t, $t \geqslant 0$, be a Brownian motion. Let

$$\Delta^n = \{s = t_0^n < t_1^n < \cdots < t_{k_n}^n = t\}, \quad n \in \mathbb{N},$$

be a sequence of partitions of the interval $[s, t]$ such that

$$|\Delta^n| = \max_i \Delta t_i^n = \max_i |t_{i+1}^n - t_i^n| \to 0, \quad n \to \infty.$$

Then,

$$\sum_{i=0}^{k_n-1} \left(\Delta B_i^n\right)^2 = \sum_{i=0}^{k_n-1} \left(B(t_{i+1}^n) - B(t_i^n)\right)^2 \overset{L^2}{\to} t - s, \quad n \to \infty.$$

REMARK 1.1.– According to this theorem:

1) We say that a Brownian motion has, on every interval $[s, t]$, the *quadratic variation* equal to $t - s$. This fact is often symbolically written as $(\mathrm{d}B_t)^2 = \mathrm{d}t$.

2) Brownian motion almost surely (with probability one) has infinite variation on every interval $[s, t]$.

DEFINITION 1.2.– We denote by $\mathcal{F}_t = \mathcal{F}_t^B$ the σ-algebra $\sigma\{B_s, s \leqslant t\}$ generated by all random variables B_s, $s \leqslant t$, and all zero-probability events. In other words, $\mathcal{F}_t$ is the smallest σ-algebra with respect to which all these random variables are measurable and which includes all zero-probability events. Then, we call the history (or past) of Brownian motion up to moment t, the class $\mathcal{H}_t$ of all random variables that are measurable with respect to the σ-algebra $\mathcal{F}_t$. Intuitively, $\mathcal{H}_t$ consists of all random variables that are completely determined by the values of Brownian motion up to the time moment t. We also denote by $\mathcal{H}_t^b$ all bounded random variables from $\mathcal{H}_t$.

A random process X_t, $t \in I \subset [0, \infty)$, is called *adapted*[3] (with respect to Brownian motion) in the interval I if $X_t \in \mathcal{H}_t$ (i.e. X_t is an $\mathcal{F}_t$-measurable random variable) for all $t \in I$.

---

3 Also called *non-anticipating*.

REMARK 1.2.– Often, it is convenient to consider general families of σ-algebras $\mathcal{F}_t$, $t \geqslant 0$, requiring the following conditions to be satisfied:

1) $\mathbb{F} = \{\mathcal{F}_t,\, t \geqslant 0\}$ is an increasing family of σ-algebras, i.e. $\mathcal{F}_s \subset \mathcal{F}_t$ for $s \leqslant t$;

2) all σ-algebras $\mathcal{F}_t$ contain zero-probability events;

3) for all $t \geqslant 0$, B_t is an $\mathcal{F}_t$-measurable random variable, and the increments $B_s - B_u$, $s \geqslant u \geqslant t$, are independent of the σ-algebra $\mathcal{F}_t$, i.e. of all $\mathcal{F}_t$-measurable random variables.

A family of σ-algebras $\mathbb{F}$ satisfying conditions (1) and (2) is called a *filtration*. Moreover, if condition (3) is satisfied, then we say that Brownian motion B is adapted to filtration $\mathbb{F}$ or that B is a Brownian motion with respect to filtration $\mathbb{F}$.

For details, see, for example, [MAC 11], Chapter 2.

1.2. Stochastic integrals

Recall that a random process $X = \{X_t,\, t \in [0, T]\}$ is called adapted (with respect to Brownian motion B) in the interval $[0, T]$ if $X_t \in \mathcal{H}_t$ for all $t \in [0, T]$.

DEFINITION 1.3.– We denote by $H^2 = H^2[0, T]$ the class of all adapted random processes X in the interval $[0, T]$, such that

$$\|X\| = \|X\|_{H^2} := \left(\mathbf{E} \int_0^T X_t^2 \mathrm{d}t \right)^{1/2} < +\infty.$$

We say that a sequence of random processes $\{X^n \in H^2\}$ converges to a random process $X \in H^2$ in the space $H^2[0, T]$ and write $X^n \overset{H^2}{\to} X$ if $\|X^n - X\| \to 0$ as $n \to \infty$.

The function $\| \cdot \|$ has the usual properties of a norm:

1) $\|\alpha X\| = |\alpha|\, \|X\|$, $\alpha \in \mathbb{R}$;

2) $\|X + Y\| \leqslant \|X\| + \|Y\|$;

3) $\big|\|X\| - \|Y\|\big| \leqslant \|X - Y\|$.

The set $H^2[0, T]$ becomes a normed and even a Banach space with norm $\|\cdot\|$ if we consider two random processes $X, \widetilde{X} \in H^2[0, T]$ as coinciding whenever $\|X - \widetilde{X}\| = 0$. It is also a Hilbert space with scalar product $(X, Y) := \mathbf{E} \int_0^T X_t Y_t \mathrm{d}t$.

DEFINITION 1.4.– An adapted random process $X = \{X_t,\, t \in [0, T]\}$ is called a *step process* if there is a partition $0 = t_0 < t_1 < \cdots < t_k = T$ of the interval $[0, T]$, such that

$$X_t = X_{t_i} \quad \text{for } t \in [t_i, t_{i+1}),\ i = 0, 1, 2, \ldots, k - 1,$$

or, shortly, $X = \sum_{i=0}^{k-1} X_{t_i} \mathbf{1}_{[t_i, t_{i+1})}$ (i.e. $X_t(\omega) = \sum_{i=0}^{k-1} X_{t_i}(\omega) \mathbf{1}_{[t_i, t_{i+1})}(t)$). The *stochastic integral* (or *Itô integral*) with respect to Brownian motion B in the interval $[0, T]$ is the sum

$$\int_0^T X_t \mathrm{d}B_t := \sum_{i=0}^{k-1} X_{t_i}\big(B_{t_{i+1}} - B_{t_i}\big).$$

We denote by $S^2[0, T]$ the class of all step processes belonging to $H^2[0, T]$ and by $S_b[0, T]$ all bounded step processes.

The stochastic integral of process $X \in S[0, T]$ with respect to a Brownian motion in subinterval $[T_1, T_2] \subset [0, T]$ is defined as

$$\int_{T_1}^{T_2} X_t \mathrm{d}B_t := \int_0^T X_t \mathbf{1}_{[T_1, T_2]}(t) \mathrm{d}B_t.$$

THEOREM 1.4.– The stochastic integral in the class $S_b = S_b[0, T]$ possesses the following properties:

1) $X, Y \in S_b,\ \alpha, \beta \in \mathbb{R}$
$\implies \int_0^T (\alpha X + \beta Y)_t \mathrm{d}B_t = \alpha \int_0^T X_t \mathrm{d}B_t + \beta \int_0^T Y_t \mathrm{d}B_t$.

2) $\mathbf{E} \int_0^T X_t \mathrm{d}B_t = 0$; if $Z \in \mathcal{H}_s$ is a bounded random variable, then $Z \int_s^T X_t \mathrm{d}B_t = \int_s^T Z X_t \mathrm{d}B_t$, and, in particular $\mathbf{E}(Z \int_s^T X_t \mathrm{d}B_t) = 0$.

3) $\mathbf{E}(\int_0^T X_t \mathrm{d}B_t)^2 = \mathbf{E}\int_0^T X_t^2 \mathrm{d}t$, i.e. $\| \int_0^T X_t \mathrm{d}B_t \|_{L^2} = \|X\|_{H^2}$

4) $\mathbf{E}(\int_0^T X_t \mathrm{d}B_t \int_0^T Y_t \mathrm{d}B_t) = \mathbf{E}\int_0^T X_t Y_t \mathrm{d}t$.

5) $\mathbf{E}(\int_0^T X_t \mathrm{d}B_t)^4 \leqslant 36\,\mathbf{E}(\int_0^T X_t^2 \mathrm{d}t)^2$.

THEOREM 1.5.– For every random process $X \in H^2[0, T]$, there is a sequence of step processes $\{X^n\}$ that converges to X in $H^2[0, T]$, i.e. such that

$$\left\| X^n - X \right\|^2 = \mathbf{E}\int_0^T \left(X_t^n - X_t \right)^2 \mathrm{d}t \to 0, \quad n \to \infty.$$

In other words, the class $S_b[0, T]$ is dense in $H^2[0, T]$.

DEFINITION 1.5.– Suppose that $X \in H^2[0, T]$ and $S_b[0, T] \ni X^n \xrightarrow{H^2} X$. The stochastic (or Itô) integral of a process X with respect to Brownian motion in the interval $[0, T]$ is the limit

$$\int_0^T X_t \mathrm{d}B_t := L^2\text{-}\lim_{n\to\infty} \int_0^T X_t^n \mathrm{d}B_t.$$

The latter limit always exists and does not depend on the choice of the sequence $S_b[0, T] \ni X^n \xrightarrow{H^2} X$.

THEOREM 1.6.– The stochastic integral in the class $H^2 = H^2[0, T]$ possesses all properties 1–5 formulated in theorem 1.4, i.e.

1) $X, Y \in H^2$, $\alpha, \beta \in \mathbb{R} \implies \int_0^T (\alpha X + \beta Y)_t \mathrm{d}B_t = \alpha \int_0^T X_t \mathrm{d}B_t + \beta \int_0^T Y_t \mathrm{d}B_t$;

2) $\mathbf{E}\int_0^T X_t \mathrm{d}B_t = 0$; for bounded $Z \in \mathcal{H}_s$, $Z \int_s^T X_t \mathrm{d}B_t = \int_s^T Z X_t \mathrm{d}B_t$, and, in particular, $\mathbf{E}(Z \int_s^T X_t \mathrm{d}B_t) = 0$;

3) $\mathbf{E}(\int_0^T X_t \mathrm{d}B_t)^2 = \mathbf{E}\int_0^T X_t^2 \mathrm{d}t$, i.e. $\| \int_0^T X_t \mathrm{d}B_t \|_{L^2} = \|X\|_{H^2}$;

4) $\mathbf{E}(\int_0^T X_t \mathrm{d}B_t \int_0^T Y_t \mathrm{d}B_t) = \mathbf{E}\int_0^T X_t Y_t \mathrm{d}t$;

5) $\mathbf{E}(\int_0^T X_t \mathrm{d}B_t)^4 \leqslant 36\, \mathbf{E}(\int_0^T X_t^2 \mathrm{d}t)^2$;

Moreover, the following Doob inequalities hold for the maximum of the stochastic integral:

(6a) $\mathbf{P}\{\sup_{t\leqslant T} |\int_0^t X_s \mathrm{d}B_s| \geqslant \lambda\} \leqslant \frac{1}{\lambda^p} \mathbf{E}|\int_0^T X_s \mathrm{d}B_s|^p$, $\lambda > 0$, $p \geqslant 1$; in particular,
$\mathbf{P}\{\sup_{t\leqslant T} |\int_0^t X_s \mathrm{d}B_s| \geqslant \lambda\} \leqslant \frac{1}{\lambda^2} \mathbf{E}\int_0^T X_t^2 \mathrm{d}t$, $\lambda > 0$;

(6b) $\mathbf{E}\sup_{t\leqslant T}(\int_0^t X_s \mathrm{d}B_s)^2 \leqslant 4\mathbf{E}\int_0^T X_t^2 \mathrm{d}t$.

The stochastic integral of a continuous adapted process can be defined similarly to Stieltjes-type integrals. It is important then, that in the Stieltjes-type integral sums, we necessarily have to take values of the integrated process at the left points of the partition intervals.

THEOREM 1.7.– Let $X \in H^2[0,T]$ be an adapted process which is L^2-continuous in the following sense: for all $t \in [0,T]$,

$$\mathbf{E}X_t^2 < \infty \qquad \text{and} \qquad \mathbf{E}|X_s - X_t|^2 \to 0 \quad \text{as} \quad s \to t.$$

If $\Delta^n = \{0 = t_0^n < t_1^n < \cdots < t_{k_n}^n = T\}$, $n \in \mathbb{N}$, is a sequence of partitions of $[0,T]$ such that $|\Delta^n| = \max_i \Delta t_i = \max_i |t_{i+1}^n - t_i^n| \to 0$, then

$$\int_0^T X_t \mathrm{d}B_t = L^2\text{-}\lim_{n\to\infty} \sum_{i=0}^{k_n-1} X(t_i^n)\big(B(t_{i+1}^n) - B(t_i^n)\big).$$

The stochastic integral can be (slightly but very usefully) extended to the random processes X for which

$$\mathbf{P}\Big\{\int_0^T X_t^2 \mathrm{d}t < +\infty\Big\} = 1. \qquad\qquad [\otimes]$$

DEFINITION 1.6.– We denote by $\widehat{H}^2[0,T]$ the class of all adapted random processes X satisfying condition $[\otimes]$.

For a process $X \in \widehat{H}^2[0,T]$, we denote

$$X_t^{(N)} = X_t \mathbf{1}_{\{\int_0^t X_s^2 \mathrm{d}s \leqslant N\}}, \quad t \in [0,T], \; N \in \mathbb{N}.$$

The stochastic (Itô) integral of X is defined by

$$\int_0^T X_t \mathrm{d}B_t := \lim_{N \to \infty} \int_0^T X_t^{(N)} \mathrm{d}B_t.$$

THEOREM 1.8.– For every $X \in \widehat{H}^2[0,T]$, the latter limit exists (with probability 1), i.e. the stochastic integral $\int_0^T X_t \mathrm{d}B_t$ is correctly defined.

THEOREM 1.9.– For every process $X \in \widehat{H}^2[0,T]$, the stochastic integral

$$I(t) := \int_0^t X_s \mathrm{d}B_s, \quad t \in [0,T],$$

is a continuous function of the upper bound t (a.s.).

We further give two more generalizations of the stochastic integrals, for *infinite* and *random* time intervals.

DEFINITION 1.7.– By $H^2[0,\infty)$, we denote the class of all adapted random processes $X = \{X_t,\ t \geqslant 0\}$ for which

$$\|X\| = \|X\|_{H^2} := \left(\mathbf{E} \int_0^\infty X_t^2 \mathrm{d}t \right)^{1/2} < +\infty.$$

We say that a sequence of adapted random processes $\{X^n\}$ converges to an adapted process X in the space $H^2[0,\infty)$ and write $X^n \xrightarrow{H^2} X$ if $\|X^n - X\| \to 0$ as $n \to \infty$.

The stochastic integral (or Itô integral) of a random process $X \in H^2[0,\infty)$ with respect to Brownian motion B in the interval $[0,\infty)$ is defined as the limit

$$\int_0^\infty X_t \mathrm{d}B_t := L^2\text{-} \lim_{n \to \infty} \int_0^n X_t \mathrm{d}B_t.$$

THEOREM 1.10.– All the properties of the stochastic integral stated in theorem 1.6 remain true with $T = \infty$.

DEFINITION 1.8.– If τ is a Markov moment (or stopping time) with respect to Brownian motion B, then we say that an adapted random process X belongs to the class $H^2[0, \tau]$ if

$$\mathbf{E} \int_0^\tau X_t^2 \mathrm{d}t = \mathbf{E} \int_0^\infty X_t^2 \mathbf{1}_{[0,\tau]}(t) \mathrm{d}t < +\infty.$$

The stochastic integral on the interval $[0, \tau]$ is defined as the random variable

$$\int_0^\tau X_t \mathrm{d}B_t := \int_0^\infty X_t \mathbf{1}_{[0,\tau]}(t) \mathrm{d}B_t.$$

THEOREM 1.11.– If τ is a Markov moment with respect to Brownian motion B, then all the properties of the stochastic integral stated in theorem 1.6 remain true with $T = \tau$.

THEOREM 1.12 (Itô formula for Brownian motion).– If $F \in C^2([0, T] \times \mathbb{R})$, then

$$F(T, B_T) - F(0, B_0) = \int_0^T F_x'(t, B_t) \mathrm{d}B_t + \int_0^T F_t'(t, B_t) \mathrm{d}t + \frac{1}{2} \int_0^T F_{xx}''(t, B_t) \mathrm{d}t.$$

For details, see, for example, [MAC 11], chapter 4.

1.3. Martingales, Itô processes and general Itô's formula

The compelling reason for studying martingales is that they pop like mushrooms all over probability theory

J. Michael Steele [STE 01]

DEFINITION 1.9.– An adapted random process $M = \{M_t, t \geqslant 0\}$ is called a *martingale* if

$$\mathbf{E}\big(Z(M_s - M_t)\big) = 0 \qquad\qquad [1.3]$$

for all $Z \in \mathcal{H}_t^b$ and $s \geqslant t \geqslant 0$.

REMARK 1.3.–   1) Random variables Z_1 and Z_2 are said to be orthogonal if $\mathbf{E}(Z_1 Z_2) = 0$. Therefore, property (1.3) can be interpreted as follows: a random process M is a martingale if the increments $M_s - M_t$, $s \geqslant t \geqslant 0$, are orthogonal to the past $\mathcal{H}_t$ or, more precisely, to all $Z \in \mathcal{H}_t^b$. Brownian motion B is a martingale – its increments are not only orthogonal to but even independent of the past. Many properties of Brownian motion can be generalized for martingales, since their proofs are often based on the orthogonality of increments to the past, rather than on their independence of the latter.

2) The requirement of boundedness of Z in equation (1.3) can often be relaxed to the integrability conditions, ensuring the integrability of the products $Z(M_s - M_t)$. For example, if $\mathbf{E} M_t^2 < \infty$, $t \geqslant 0$, then from equation (1.3) for all $Z \in \mathcal{H}_t^b$, it follows that equation (1.3) also holds for all $Z \in \mathcal{H}_t$ with a finite second moment ($\mathbf{E} Z^2 < \infty$).

3) Most often, a martingale is defined using the notion of the conditional expectation of a random variable with respect to a σ-algebra. Let X be a random variable, and $\mathcal{G} \subset \mathcal{F}$ a σ-algebra. A random variable Y is called the expectation of X with respect to the σ-algebra $\mathcal{G}$ if it is $\mathcal{G}$-measurable and

$$\mathbf{E}(ZX) = \mathbf{E}(ZY)$$

for every $\mathcal{G}$-measurable bounded random variable Z. It is denoted by $\mathbf{E}(X|\mathcal{G})$. Then, property [1.3] defining a martingale can be written as follows:

$$\mathbf{E}(M_s|\mathcal{F}_t) = M_t, \quad s \geqslant t \geqslant 0.$$

4) Calculation of conditional expectations with respect to a σ-algebra, in general, is not an easy task. However, in practical calculations, it can often be reduced to calculation of expectations of the particular form

$$\mathbf{E}(f(X,Y)|Y) := \mathbf{E}(f(X,Y)|\sigma(Y)),$$

where $f\colon \mathbb{R}^2 \to \mathbb{R}$ is a "good" (measurable) function, X and Y are independent random variables and $\sigma(Y)$ is the σ-algebra generated by Y. Under appropriate integrability conditions (not very restrictive), we have the following intuitively clear formula:

$$\mathbf{E}(f(X,Y)|Y) = \mathbf{E}f(X,y)\Big|_{y=Y}, \qquad [1.4]$$

i.e. we have to calculate the expectation $g(y) := \mathbf{E}f(X, y)$ for every fixed $y \in \mathbb{R}$ and then take $y = Y$ in the obtained function $g(y)$.

THEOREM 1.13.– Let $H \in H^2[0, T]$. Denote $M_t = \int_0^t H_s dB_s$ and $\langle M \rangle_t = \int_0^t H_s^2 ds$, $t \in [0, T]$. Then:

1) the random process M_t, $t \in [0, T]$, is a martingale;

2) $\mathbf{E}(Z(M_s^2 - M_t^2)) = \mathbf{E}(Z(M_s - M_t)^2)$ for all $Z \in \mathcal{H}_t^b$, $0 \leqslant t \leqslant s \leqslant T$;

3) the random process $N_t := M_t^2 - \langle M \rangle_t$, $t \in [0, T]$, is a martingale.

DEFINITION 1.10.– The stochastic integral of a random process Y_s, $s \in [0, T]$, with respect to the martingale M defined in proposition 1.13 is the random variable

$$\int_0^T Y_s dM_s := \int_0^T Y_s H_s dB_s,$$

provided that the integral on the right-hand side is defined.

The definition is justified by the statements on the convergence of Riemann-type integral sums. Let $\Delta^n = \{0 = t_0^n < t_1^n < \cdots < t_{k_n}^n = T\}$, $n \in \mathbb{N}$, be a sequence of partitions of the interval $[0, T]$ with $|\Delta^n| = \max_i |t_{i+1}^n - t_i^n| \to 0$, $n \to \infty$. To simplify the notation, we omit the indices n and write $Y_i = Y(t_i^n)$, $\Delta Y_i = Y_{i+1} - Y_i$.

THEOREM 1.14.– Let $H \in H^2[0, T]$, $M_t := \int_0^t H_s dB_s$, $t \in [0, T]$, and $\langle M \rangle_t := \int_0^t H_s^2 ds$, $t \in [0, T]$. If Y_t, $t \in [0, T]$, is a continuous adapted process, then:

1) $\sum_i Y_i \Delta M_i \overset{\mathbf{P}}{\to} \int_0^T Y_t dM_t = \int_0^T Y_t H_t dB_t$, $n \to \infty$;

2) $\sum_i Y_i \Delta M_i^2 \overset{\mathbf{P}}{\to} \int_0^T Y_t d\langle M \rangle_t = \int_0^T Y_t H_t^2 dt$, $n \to \infty$.
In particular, by taking $Y \equiv 1$, we have

3) $\sum_i \Delta M_i^2 \overset{\mathbf{P}}{\to} \langle M \rangle_T$, $n \to \infty$.
Therefore, the random process $\langle M \rangle_t$, $t \in [0, T]$, is called the quadratic variation of the martingale M.

DEFINITION 1.11.– An adapted random process X_t, $t \in [0, T]$, is called an Itô process (or diffusion-type process) if it can be written in the form

$$X_t = X_0 + \int_0^t K_s \mathrm{d}s + \int_0^t H_s \mathrm{d}B_s, \quad t \in [0, T], \tag{1.5}$$

where K and H are adapted random processes such that the integrals on the right-hand side exist (a.s.), i.e. $\int_0^T |K_s| \mathrm{d}s < +\infty$ and $\int_0^T H_s^2 \mathrm{d}s < +\infty$ (a.s.). In this case, we say that the process X admits the stochastic differential

$$\mathrm{d}X_t = K_t \mathrm{d}t + H_t \mathrm{d}B_t,$$

or, shortly,

$$dX = K\mathrm{d}t + H\mathrm{d}B.$$

In particular, when $H = 0$, we shall say that the Itô process X is regular.

The stochastic integral of an adapted random process Z_t, $t \in [0, T]$, with respect to the Itô process X as in equation [1.5] is the stochastic process

$$Z \bullet X_t = \int_0^t Z_s \mathrm{d}X_s := \int_0^t Z_s K_s \mathrm{d}s + \int_0^t Z_s H_s \mathrm{d}B_s, \quad t \in [0, T],$$

provided that the integrals on the right-hand side exist for all $t \in [0, T]$. Note that, in such a case, the stochastic integral $Z \bullet X$ is also an Itô process.

DEFINITION 1.12.– The *covariation* of two Itô processes

$$X_t = X_0 + \int_0^t K_s \mathrm{d}s + \int_0^t H_s \mathrm{d}B_s \text{ and } Y_t = Y_0 + \int_0^t \widetilde{K}_s \mathrm{d}s + \int_0^t \widetilde{H}_s \mathrm{d}B_s$$

is the random process $\langle X, Y \rangle_t := \int_0^t H_s \widetilde{H}_s \, ds$. The process $\langle X \rangle_t := \langle X, X \rangle_t$ is called the quadratic variation of the process X. Note that the covariation $\langle X, Y \rangle = 0$ if at least one of the processes X and Y is regular.

THEOREM 1.15 (Properties of the stochastic integral).– Let X and Y be Itô processes, and let Z and W be adapted random processes in the interval $[0, T]$. Then:

1) $Z \bullet (\alpha X + \beta Y) = \alpha Z \bullet X + \beta Z \bullet Y, \alpha, \beta \in \mathbb{R}$;

2) $(\alpha Z + \beta W) \bullet X = \alpha Z \bullet X + \beta W \bullet X, \alpha, \beta \in \mathbb{R}$;

3) $W \bullet (Z \bullet X) = (WZ) \bullet X$;

4) $\langle Z \bullet X, W \bullet Y \rangle = (ZW) \bullet \langle X, Y \rangle$,

provided that the integrals on the right-hand sides of the first three equalities and on the left-hand side of the last equality are well defined.

REMARK 1.4.– When calculating stochastic integrals (and ordinary integrals as well), it is often convenient to apply formal rules of summation and multiplication of random processes and their differentials. We shall write $dZ = Y dX$ if $Z_t - Z_0 = \int_0^t Y_s dX_s, t \in [0, T]$. In particular (when $Y = 1$), for two Itô processes Z and X, we shall write $dZ = dX$ if $Z_t - Z_0 = X_t - X_0, t \in [0, T]$. We also define the sum and the product of the differentials of Itô processes X and Y by

$$dX + dY := d(X + Y), \qquad dX \cdot dY := d\langle X, Y \rangle.$$

Then, the properties stated in proposition 1.15 can be written in the differential form as follows:

1) $Zd(\alpha X + \beta Y) = \alpha Z dX + \beta Z dY$,

2) $(\alpha Z + \beta W) dX = \alpha Z dX + \beta W dX$,

3) $W(Z dX) = (WZ) dX$,

4) $(Z dX) \cdot (W dY) = (ZW) dX \cdot dY$.

Also note the following properties that directly follow from the definition:

5) $(dX + dY) \cdot dZ = dX \cdot dZ + dY \cdot dZ$,

6) $dX \cdot dY = dY \cdot dX$,

7) $(dX \cdot dY) \cdot dZ = dX \cdot (dY \cdot dZ) = 0$, since the covariation of two Itô processes is always a regular process.

Consider, say, property (3). Let $Y_t = Y_0 + \int_0^t Z_s dX_s$ or, in the differential form, $dY = Z dX$. By property (3), we may formally multiply this equality by a random process W, thus obtaining the equality $W dY = W Z dX$; then, formally integrating this, we obtain the correct equality $\int_0^t W_s dY_s = \int_0^t W_s Z_s dX_s$.

We now state a generalization of theorem 1.15 for Itô processes.

THEOREM 1.16.– Let $X_t = X_0 + A_t + M_t = X_0 + \int_0^t K_s ds + \int_0^t H_s dB_s$, $t \in [0, T]$, be an Itô process, and let Y_t, $t \in [0, T]$, be a continuous adapted random process. Then, we have:

(1) $\sum_i Y_i \Delta X_i \overset{\mathbf{P}}{\to} \int_0^T Y_t dX_t, \quad n \to \infty$;

(2) $\sum_i Y_i \Delta X_i^2 \overset{\mathbf{P}}{\to} \int_0^T Y_t d\langle X \rangle_t, \quad n \to \infty$.

In particular, for $Y \equiv 1$, we have

$$\sum_i \Delta X_i^2 \overset{\mathbf{P}}{\to} < X >_T,\ n \to \infty.$$

Therefore, the random process $\langle X \rangle_t$, $t \in [0, T]$, is called the quadratic variation of X.

COROLLARY 1.2.– If X_t and Y_t, $t \in [0, T]$, are two Itô processes, then

$$\sum_i \Delta X_i \Delta Y_i \overset{\mathbf{P}}{\to} \langle X, Y \rangle_T, \quad n \to \infty.$$

THEOREM 1.17 (Itô's formula for Itô processes).– If $X_t = X_0 + \int_0^t K_s ds + \int_0^t H_s dB_s$, $t \in [0, T]$, is an Itô process and $F \in C^2([0, T] \times \mathbb{R})$, then

$$F(T, X_T) - F(0, X_0)$$

$$= \int_0^T F_x'(t, X_t) dX_t + \int_0^T F_t'(t, X_t) dt + \frac{1}{2} \int_0^T F_{xx}''(t, X_t) d\langle X \rangle_t$$

$$= \int_0^T F_x'(t, X_t) H_t dB_t + \int_0^T F_x'(t, X_t) K_t dt$$

$$+ \int_0^T F_t'(t, X_t) dt + \frac{1}{2} \int_0^T F_{xx}''(t, X_t) H_t^2 dt.$$

THEOREM 1.18 (Integration-by-parts formula).– If X and Y are Itô processes in the time interval $[0, T]$, then

$$X_t Y_t = X_0 Y_0 + \int_0^t X_s dY_s + \int_0^t Y_s dX_s + \langle X, Y \rangle_t, \quad t \in [0, T],$$

or, in a differential form,

$$d(XY) = X dY + Y dX + d\langle X, Y \rangle.$$

In particular, if at least one of the processes X and Y is regular, these formulas become the usual integration-by-parts formulas

$$\int_0^t X_s dY_s = X_t Y_t - X_0 Y_0 - \int_0^t Y_s dX_s, \qquad X dY = d(XY) - Y dX.$$

REMARK 1.5.– Theorems 1.17 and 1.18 remain true if T is a finite (a.s.) Markov moment.

For details, see, for example, Chapter 7 of [MAC 11].

1.4. Stochastic differential equations

DEFINITION 1.13.– A continuous random process X_t, $t \in I \subset [0, \infty)$, is a solution of the stochastic differential equation

$$dX_t = b(X_t, t)dt + \sigma(X_t, t)dB_t, \quad X_0 = x_0, \tag{1.6}$$

in an interval I if, for all $t \in I$, it satisfies a.s. the equation

$$X_t = x_0 + \int_0^t b(X_s, s)\mathrm{d}s + \int_0^t \sigma(X_s, s)\mathrm{d}B_s, \quad t \in I. \qquad [1.7]$$

Both equations are called stochastic differential equations (SDEs), though here the word *integral* would fit better.

EXAMPLE 1.1.– Let us check that the random process

$$X_t = x_0 \exp\{(\mu - \sigma^2/2)t + \sigma B_t\}, \quad t \geqslant 0,$$

is a solution of the SDE

$$\mathrm{d}X_t = \mu X_t \mathrm{d}t + \sigma X_t \mathrm{d}B_t, \quad X_0 = x_0.$$

This equation describes, for example, the dynamics of stock price in financial mathematics (Black–Scholes model). Applying Itô's (theorem 1.12) to the function $F(t, x) = x_0 \exp\left\{(\mu - \sigma^2/2)t + \sigma x\right\}$, we get

$$X_t = F(t, B_t)$$

$$= F(0,0) + \int_0^t F_x'(s, B_s)\mathrm{d}B_s + \int_0^t F_s'(s, B_s)\mathrm{d}s + \frac{1}{2}\int_0^t F_{xx}''(s, B_s)\mathrm{d}s$$

$$= x_0 + \int_0^t x_0\sigma \exp\left\{(\mu - \sigma^2/2)s + \sigma B_s\right\}\mathrm{d}B_s$$

$$+ \int_0^t x_0(\mu - \sigma^2/2) \exp\left\{(\mu - \sigma^2/2)s + \sigma B_s\right\}\mathrm{d}s$$

$$+ \frac{1}{2}\int_0^t x_0\sigma^2 \exp\left\{(\mu - \sigma^2/2)s + \sigma B_s\right\}\mathrm{d}s$$

$$= x_0 + \int_0^t \mu x_0 \exp\left\{(\mu - \sigma^2/2)s + \sigma B_s\right\}\mathrm{d}s$$

$$+ \int\limits_0^t \sigma x_0 \exp\left\{(\mu - \sigma^2/2)s + \sigma B_s\right\} \mathrm{d}B_s$$

$$= x_0 + \int\limits_0^t \mu X_s \mathrm{d}s + \int\limits_0^t \sigma X_s \mathrm{d}B_s, \quad t \geqslant 0.$$

In particular, for $\mu = 0$ and $\sigma = 1$, we get that the random process

$$X_t = x_0 \exp\{B_t - t/2\}, \quad t \geqslant 0,$$

is a solution of the stochastic differential equation

$$\mathrm{d}X_t = X_t \mathrm{d}B_t, \qquad X_0 = x_0.$$

Therefore, the process X_t is sometimes called the stochastic exponent of a Brownian motion. (If B were not a Brownian motion but a continuously differentiable process, the exponent $X_t = x_0 e^{B_t}$ would be a solution of the equation.)

We will further introduce the following Lipschitz and linear growth conditions for the coefficients:

(1) $|b(x,t) - b(y,t)| + |\sigma(x,t) - \sigma(y,t)| \leqslant L|x - y|$, $x, y \in \mathbb{R}$, $t \in I$;

(2) $|b(x,t)| + |\sigma(x,t)| \leqslant L(1 + |x|)$, $x \in \mathbb{R}$, $t \in I$.

THEOREM 1.19.– If conditions (1) and (2) are satisfied, then there exists a unique continuous process X that satisfies the stochastic differential equation [1.6] in the interval I.

REMARK 1.6.– Under the conditions of the theorem, the solution of [1.6] is a Markov process. This means that the conditional distribution of X_s, $s > t$, given the history $\mathcal{H}_t$ (or, equivalently, σ-algebra $\mathcal{F}_t$), is the same as the conditional distribution of X_s given only X_t. More precisely,

$$\mathbf{P}\{X_s < x | \mathcal{F}_t\} = \mathbf{P}\{X_s < x | X_t\}, \quad x \in \mathbb{R}, \ s > t.$$

Intuitively, this means that the statistical behavior of the process in the "future" (the values of X_s at time moments $s > t$), given the "past" $\mathcal{H}_t$, depends, in fact, only on the "present" state X_t of the process.

DEFINITION 1.14.– Let X^x denote the solution of a time-homogeneous SDE (the coefficients do not depend on t)

$$dX_t = b(X_t)\mathrm{d}t + \sigma(X_t)\mathrm{d}B_t \qquad\qquad [1.8]$$

with the initial condition $X_0 = x$. The family of processes $X = \{X^x, x \in \mathbb{R}\}$ is called a (homogeneous) diffusion process. The *generator* (or *infinitesimal operator*) of diffusion process $X = \{X^x\}$ is the operator A in a set of functions $f : \mathbb{R} \to \mathbb{R}$ defined by

$$Af(x) = \lim_{t\downarrow 0} \frac{T_t f(x) - f(x)}{t} = \lim_{t\downarrow 0} \frac{\mathbf{E}f(X_t^x) - f(x)}{t}, \quad x \in \mathbb{R}.$$

The definition domain $\mathcal{D}_A$ of the generator A consists of all functions $f : \mathbb{R} \to \mathbb{R}$ for which the latter (finite) limit exists for all $x \in \mathbb{R}$.

THEOREM 1.20.– Let $X = \{X^x\}$ be a homogeneous diffusion process defined by equation [1.8] with Lipschitz coefficients. If a function $f \in C^2(\mathbb{R})$, together with its first- and second-order derivatives, is of polynomial growth,[4] then $f \in \mathcal{D}_A$, and

$$Af(x) = b(x)f'(x) + \frac{1}{2}\sigma^2(x)f''(x), \quad x \in \mathbb{R},$$

or, in short, $Af = bf' + \frac{1}{2}\sigma^2 f''$. Moreover, for every such function f,

$$\mathbf{E}f(X_t^x) = f(x) + \mathbf{E}\int_0^t Af(X_s^x)\mathrm{d}s, \quad t \geqslant 0,$$

or

$$T_t f(x) = f(x) + \int_0^t T_s Af(x)\mathrm{d}s, \quad t \geqslant 0 \quad \text{(Dynkin's formula)}.$$

---

4 That is, $|f^{(i)}(x)| \leqslant C(1 + |x|^p)$, $x \in \mathbb{R}$, $i = 0, 1, 2$, for some $p \in \mathbb{N}$.

COROLLARY 1.3.– (1) *(Backward Kolmogorov equation)* Suppose that $X = \{X^x\}$ is a homogeneous diffusion process with generator A. For $f \in C_b^2(\mathbb{R})$, denote

$$u(t, x) = T_t f(x) = \mathbf{E} f(X_t^x), \quad (t, x) \in \mathbb{R}_+ \times \mathbb{R}.$$

Then, the function u is a solution of the partial differential equation with the initial condition

$$\begin{cases} \frac{\partial u}{\partial t} = Au, \\ u(0, x) = f(x), \end{cases}$$

where the operator A is taken with respect to the argument x of the function $u = u(t, x)$.

(2) *(Feynman–Kac formula)* If, in addition, $g \in C_b(\mathbb{R})$, then the function

$$v(t, x) := \widetilde{\mathbf{E}} \left[\exp \left\{ \int_0^t g(X_s^x) \, ds \right\} f(X_t^x) \right]$$

is a solution of the partial differential equation with the initial condition

$$\begin{cases} \frac{\partial v}{\partial t} = Au - gv, \\ v(0, x) = f(x). \end{cases}$$

The notion of a stochastic differential equation can also be generalized to the case where the stochastic integral is taken with respect to an arbitrary Itô process.

DEFINITION 1.15.– Let Z_t, $t \geqslant 0$, be an Itô process, and A_t, $t \geqslant 0$, a regular Itô process. A random (Itô) process X_t, $t \geqslant 0$, is said to be a solution of the stochastic differential equation

$$dX_t = b(X_t, t) \, dA_t + \sigma(X_t, t) \, dZ_t, \qquad X_0 = x_0, \qquad [1.9]$$

with coefficients $b, \sigma \colon \mathbb{R} \times [0, \infty) \to \mathbb{R}$ if it satisfies the equation

$$X_t = x_0 + \int_0^t b(X_s, s) \, dA_s + \int_0^t \sigma(X_s, s) \, dZ_s$$

for all $t \geqslant 0$.

THEOREM 1.21.– If the coefficients b and σ satisfy the Lipschitz and linear growth conditions, then equation (1.9) has a unique solution.

EXAMPLE 1.2.– Consider the stochastic differential equation

$$\mathrm{d}X_t = X_t\,\mathrm{d}Y_t, \quad X_0 = x, \tag{1.10}$$

where Y is an arbitrary Itô process. Then, the process

$$X_t = x\exp\{Y_t - \langle Y\rangle_t\}, \quad t \geqslant 0,$$

is its solution. Indeed, denoting $Z_t := Y_t - \langle Y\rangle_t$ and applying Itô's formula (theorem 1.17) to $X_t = F(Z_t)$ with $F(z) = xe^z$, we have

$$X_t = xe^{Z_t} = x + \int_0^t xe^{Z_s}\,\mathrm{d}Z_s + \frac{1}{2}\int_0^t xe^{Z_s}\,\mathrm{d}\langle Z\rangle_s$$

$$= x + \int_0^t xe^{Z_s}\,\mathrm{d}Y_s - \frac{1}{2}\int_0^t xe^{Z_s}\,\mathrm{d}\langle Y\rangle_s + \frac{1}{2}\int_0^t xe^{Z_s}\,\mathrm{d}\langle Y\rangle_s$$

$$= x + \int_0^t xe^{Z_s}\,\mathrm{d}Y_s = x + \int_0^t X_s\,\mathrm{d}Y_s, \quad t \geqslant 0,$$

since $\langle Z\rangle = \langle Y\rangle$.

The process $\mathcal{E}(Y)_t = \exp\{Y_t - \langle Y\rangle_t\}$ is called the *stochastic exponential* of an Itô process Y, since in the case of regular Itô process Y, the solution $X_t = \exp\{Y_t\}$ of the SDE [1.10] with $x = 1$ is the "usual" exponential of Y.

For details, see, for example, [MAC 11], chapters 6 and 7, [STE 01], chapter 4.

1.5. Change of probability: the Girsanov theorem

$(\Omega, \mathcal{F}, \mathbf{P})$ with some fixed probability $\mathbf{P}$. Now, we need to consider other probabilities on the same σ-algebra $\mathcal{F}$. Moreover, when passing to other probabilities, we will need to know how to "recognize" a Brownian motion with respect to these probabilities. We know that a Brownian motion is a martingale with quadratic variation $\langle B\rangle_t = t$.[5] It appears that the converse is true; it is the so-called Lévy characterization of a Brownian motion.

---

5 Defined as the limit of the sums of squared increments as the mesh of a partition of the interval $[0, t]$ tends to zero.

THEOREM 1.22 (Lévy).– A continuous martingale M_t, $t \in [0, T]$, is a Brownian motion in the interval $[0, T]$ if and only if its quadratic variation $\langle M \rangle_t = t$, $t \in [0, T]$.

Let an adapted random process λ_t, $t \in [0, T]$, satisfy the integrability condition $\int_0^T \lambda_t^2 \, dt < \infty$ (a.s.). Define the random process

$$M_t := \exp \left\{ \int_0^t \lambda_s \, dB_s - \frac{1}{2} \int_0^t \lambda_s^2 \, ds \right\}, \quad t \in [0, T]. \tag{1.11}$$

It is a solution of the stochastic differential equation (see example 1.2)

$$M_t = 1 + \int_0^t M_s \lambda_s \, dB_s. \tag{1.12}$$

If in equation [1.12], the integrand process $M\lambda \in H^2[0, T]$ (e.g. if λ is a bounded process), then M is a martingale (theorem 1.13), and $\mathbf{E}M_t = 1$, $t \in [0, T]$. This is not the case in general. However, a rather general sufficient condition is known:

THEOREM 1.23 (Novikov theorem).– If

$$\mathbf{E} \left[\exp \left(\frac{1}{2} \int_0^T \lambda_t^2 \, dt \right) \right] < \infty, \tag{1.13}$$

then the process M defined by equation [1.11] is a martingale, and $\mathbf{E}M_T = 1$.

THEOREM 1.24 (Girsanov theorem).– Suppose that the process M defined by equation [1.11] is a martingale (e.g. the Novikov condition [1.13] is satisfied). On the measurable space $(\Omega, \mathcal{F})$, define the new probability $\widetilde{\mathbf{P}}$ by

$$\widetilde{\mathbf{P}}(A) = \mathbf{E}(\mathbf{1}_A M_T), \quad A \in \mathcal{F}. \tag{1.14}$$

Then, the random process (called a Brownian motion with drift)

$$\widetilde{B}_t := B_t - \int_0^t \lambda_s \, ds, \quad t \in [0, T], \tag{1.15}$$

is a Brownian motion with respect to probability $\widetilde{\mathbf{P}}$.

REMARK 1.7.– From equation [1.14] in a usual way it follows that for every random variable $X \in \mathcal{F}_T$,

$$\widetilde{\mathbf{E}}(X) = \mathbf{E}(X M_T) \quad \text{and, conversely,} \quad \mathbf{E}(X) = \widetilde{\mathbf{E}}(X M_T^{-1}),$$

provided that the expectations on the left- or right-hand sides are finite. Here, $\widetilde{\mathbf{E}}$ denotes the expectation with respect to $\widetilde{\mathbf{P}}$.

PROOF.– We first check that the process $\widetilde{B}$ is a martingale with respect to $\widetilde{\mathbf{P}}$, i.e. for every $Z \in \mathcal{H}_s^b$,

$$\widetilde{\mathbf{E}}\big(Z(\widetilde{B}_t - \widetilde{B}_s)\big) = \mathbf{E}\left[Z\left(B_t - B_s - \int_s^t \lambda_u \, du\right) M_T\right] = 0.$$

We have

$$\mathbf{E}\left[Z\left(B_t - B_s - \int_s^t \lambda_u \, du\right) M_T\right]$$

$$= \mathbf{E}\left[Z\left(B_t - B_s - \int_s^t \lambda_u \, du\right)\left(M_s + \int_s^T M_u \lambda_u \, dB_u\right)\right]$$

$$= \mathbf{E}[Z M_s(B_t - B_s)] + \mathbf{E}\left[Z(B_t - B_s)\left(\int_s^t + \int_t^T\right) M_u \lambda_u \, dB_u\right]$$

$$\quad - \mathbf{E}\left(Z M_s \int_s^t \lambda_u \, du\right) - \mathbf{E}\left[Z\left(\int_s^t \lambda_u \, du \left(\int_s^t + \int_t^T\right) M_u \lambda_u \, dB_u\right)\right]$$

$$= 0 + \mathbf{E}\left[Z \int_s^t M_u \lambda_u \, du\right] - \mathbf{E}\left(Z M_s \int_s^t \lambda_u \, du\right)$$

$$\quad - \mathbf{E}\left[Z\left(\int_s^t \lambda_u \, du \int_s^t M_u \lambda_u \, dB_u\right)\right]$$

$$= \mathbf{E}\left[Z \int_s^t (M_u - M_s)\lambda_u \, du\right] - \mathbf{E}\left[Z \int_s^t \lambda_u \, du \, (M_t - M_s)\right]$$

$$= -\mathbf{E}\left[Z \int_s^t \int_s^u \lambda_v \, dv \, dM_u\right] \qquad \text{[Theorem 1.18]}$$

$$= -\mathbf{E}\left[Z \int_s^t \int_s^u \lambda_v \, dv \, M_v \lambda_v \, dB_u\right] = 0.$$

The quadratic variations of processes $\widetilde{B}$ and B coincide $\mathbf{P}$-a.s. As zero-probability events for both $\widetilde{\mathbf{P}}$ and $\mathbf{P}$ are the same, the quadratic variations of processes $\widetilde{B}$ and B also coincide $\widetilde{\mathbf{P}}$-a.s. Thus, $\langle \widetilde{B} \rangle_t = t$ $\widetilde{\mathbf{P}}$-a.s. Therefore, by theorem 1.22, $\widetilde{B}$ is a Brownian motion with respect to $\widetilde{\mathbf{P}}$.   □

In the simplest and most often used case where $\lambda_t = \lambda$ (a constant) for all $t \in [0, T]$, we get the following:

COROLLARY 1.4.– Let

$$M_t := \exp\{\lambda B_t - \lambda^2 t/2\}, \quad t \in [0, T].$$

On the measurable space $(\Omega, \mathcal{F})$, define the probability $\widetilde{\mathbf{P}}$ by

$$\widetilde{\mathbf{P}}(A) = \mathbf{E}(\mathbf{1}_A M_T), \quad A \in \mathcal{F}.$$

Then, the Brownian motion with linear drift,

$$\widetilde{B}_t := B_t - \lambda t, \quad t \in [0, T],$$

is a Brownian motion with respect to $\widetilde{\mathbf{P}}$. Moreover, for every random variable $X \in \mathcal{F}_T$,

$$\widetilde{\mathbf{E}}(X) = \mathbf{E}(X e^{\lambda B_T - \lambda^2 T/2}) \quad \text{and} \quad \mathbf{E}(X) = \widetilde{\mathbf{E}}(X e^{-\lambda B_T + \lambda^2 T/2}),$$

provided that the expectations on the left- or right-hand side are finite.

EXAMPLE 1.3.– The Girsanov theorem allows us, via change of probability, to make martingales (e.g. Brownian motion) with drift just martingales. Suppose that we want to calculate the expectation of a random variable that is a functional of Brownian motion with drift,

$$\mathbf{E}\big[f\big(\max_{t \leqslant T}(B_t - \lambda t)\big)\big].$$

By Corollary 1.4,

$$\mathbf{E}\big[f\big(\max_{t \leqslant T}(B_t - \lambda t)\big)\big] = \widetilde{\mathbf{E}}\big[f\big(\max_{t \leqslant T}(B_t - \lambda t)\big)e^{-\lambda B_T + \lambda^2 T/2}\big]$$

$$= \widetilde{\mathbf{E}}\big[f\big(\max_{t \leqslant T} \widetilde{B}_t\big)e^{-\lambda(\widetilde{B}_T + \lambda T) + \lambda^2 T/2}\big]$$

$$= \widetilde{\mathbf{E}}\big[f\big(\max_{t\leqslant T}\widetilde{B}_t\big)e^{-\lambda\widetilde{B}_T-\lambda^2 T/2}\big]$$

$$= \mathbf{E}\big[f\big(\max_{t\leqslant T}B_t\big)e^{-\lambda B_T-\lambda^2 T/2}\big].$$

In the last equality, we dropped the tilde, since $\widetilde{B}$ is a Brownian motion with respect to $\widetilde{\mathbf{P}}$ (as B with respect to $\mathbf{P}$). We can calculate the latter expectation, provided that we know the joint density $p(y, x)$ of the maximum of Brownian motion $B_T^* := \max_{t\leqslant T} B_t$ and its value B_T.[6] Thus,

$$E\big[f\big(\max_{t\leqslant T}(B_t-\lambda t))\big] = \mathbf{E}\big[f\big(\max_{t\leqslant T}B_t\big)e^{-\lambda B_T-\lambda^2 T/2}\big]$$

$$= \iint_{\mathbb{R}^2} f(y)\, e^{-\lambda x-\lambda^2 T/2}p(y, x)\,\mathrm{d}x\,\mathrm{d}y$$

$$= \int_0^\infty \mathrm{d}y \int_{-\infty}^y f(y)p(y, x)e^{-\lambda x-\lambda^2 T/2}\,\mathrm{d}x.$$

REMARK 1.8.– We easily see that these arguments can be applied in a more general situation. Suppose that we know the joint density $p(y, x)$ of some functional Y of a Brownian motion in the interval $[0, T]$ and of Brownian motion value itself, B_T. Then, the joint density of the same functional of a Brownian motion $\widetilde{B}$ with drift and its value $\widetilde{B}_T$ is obtained by multiplying $p(y, x)$ by $e^{-\lambda x-\lambda^2 T/2}$.

THEOREM 1.25 (Martingale representation by a stochastic integral).– Let $M_t, t \in [0, T]$, be a martingale. Then, there exists a process $H \in \widehat{H}^2[0, T]$ such that

$$M_t = M_0 + \int_0^t H_s\mathrm{d}B_s, \quad t \in [0, T].$$

Moreover, if M is a square-integrable martingale, i.e. $\mathbf{E}M_T^2 < +\infty$, then $H \in H^2[0, T]$.

---

6 It is calculated in corollary 1.1: $p(y, x) = \frac{2(2y-x)}{\sqrt{2\pi T^3}}\, e^{-\frac{(2y-x)^2}{2T}}$, $y \geqslant x^+ = \max\{x, 0\}$.

COROLLARY 1.5.– If $X \in \mathcal{H}_T$ is a square-integrable random variable (i.e. $\mathbf{E}X^2 < +\infty$), then it is representable in the form

$$X = \mathbf{E}X + \int_0^T H_s \mathrm{d}B_s$$

with $H \in H^2[0, T]$.

For details, see Chapter 13 in [STE 01].

2

The Black–Scholes Model

> I do not claim to know any geometry, but I do claim to understand quite
> clearly what geometry is.   *G.H. Hardy*

2.1. Introduction: what is an option?

An option is a contract between two parties, buyer and seller, which gives
one party the right, but not the obligation, to buy or to sell some asset until an
agreed date, while the second party has the obligation to fulfill the contract if
requested.

Let us illustrate this notion by using a real-life example. Suppose that you
have decided to buy a house and have found somewhere you like. At the
moment, you do not have enough money, but you hope to in the near future
(say, by getting credit or by selling another property). Therefore, you make a
contract (called an option) with the seller of the house which states that he
will wait for half a year and will sell the house for an agreed price of, say,
$100,000. However, he agrees to wait only on the condition that you pay for
this, say, $1,000. We can then imagine the following two scenarios:

1. Within half a year, the prices of real estate have increased, and the
market price of the house has increased to $120,000. As the owner has signed
a contract with you and will be paid for this (by selling the option), he is
obliged to sell you the house for $100,000. Thus, in this case, you have
a significant profit of

$$120,000 - 100,000 - 1,000 = \$19,000.$$

2. Checking the details of the house and talking to the neighbors, you learn that there are many dysfunctional neighbors and that the house has a very expensive and time-consuming solid-fuel heating boiler. So, although initially you thought you had found a dream house, now you clearly see that it is not worth the agreed price. Happily, according to the contract, you are not obliged to buy the house. Of course, you lose the sum of $1,000 paid for the contract.

This example illustrates some important properties of an option. First, in buying an option, you gain the right, but not the obligation, to buy something. If you do not exercise the option after the expiration time of the contract, it becomes worthless, and you lose only the sum paid for the option. Second, the option is a contract giving the right to a certain asset. Therefore, an option is a *derivative* financial instrument, whose price depends on the price of some other asset. In our example, this asset is a house. In the options used by investors, the role of such an asset is usually played by stocks or market indices.

The two main types of options are the following:

– *A call option* gives its owner the right to buy some asset (commonly a stock, a bond, currency, etc.) for a price called the *exercise price* or *strike price* until a specified time, called the *maturity* or *expiration time* of the option. If, at the expiration time, the option is not exercised (i.e. the owner decides not to buy the underlying asset), it becomes void and worthless. Call options are similar to long-term stock position, as the buyer hopes that the price of the underlying stock will significantly increase up to the expiration time.

– *A put option* gives its owner the right to sell some asset for an exercise price until the expiration time of the option. Put options are similar to short-term stock position, as the buyer hopes that the price of the underlying stock will significantly decrease until the expiration time.

The option price of an option (not to be confused with the price of underlying stock) is called the *premium*. The premium depends on factors such as the price of the underlying stock, the exercise price, the time remaining until the expiration time and the *volatility* of the stock. Determining the theoretical fair value (premium) for a call or put option, the so-called *option pricing*, is rather complicated and is the main problem considered in this course.

There are two main types of options:

– *European options*: they can only be exercised at the expiration time.

– *American options*: in contrast to European options, they can be exercised at any time before the expiration time.

Let S_t be the stock price at time moment t. The exercise price of the European call option with exercise price K and expiration time T, i.e. the profit of the option buyer, is $S_T - K$ if $S_T > K$, and zero if $S_T \leqslant K$. Thus, the profit of the call option buyer is

$$f(S_T) = (S_T - K)^+ = \begin{cases} S_T - K \text{ if } S_T > K; \\ 0 \qquad \text{if } S_T \leqslant K. \end{cases}$$

Similarly, the profit of a buyer of the European put option with exercise price K and expiration time T is

$$g(S_T) = (K - S_T)^+ = \begin{cases} K - S_T \text{ if } S_T < K; \\ 0, \qquad \text{if } S_T \geqslant K. \end{cases}$$

The following two questions arise at the time of purchasing an option:

1) What is the option premium to be paid by the buyer to the seller of, say, a call option? Or, in other words, how can we evaluate, at the time of signing the contract ($t = 0$), the future profit of the buyer $(S_T - K)^+$ at time moment T? This is the so-called problem of option pricing.

2) How can the seller of an option, after getting the premium at the moment $t = 0$, guarantee the profit $(S_T - K)^+$ at time moment T? This is the problem of option risk management or hedging.

An answer to these closely related questions is based on modeling stock prices (usually by stochastic differential equations) and working on the assumption of no-arbitrage in the financial market, which essentially means that there is no guaranteed profit without any risk.

See also the Introduction in [LAM 96] and sections 4.1.1–3 in [MIK 99].

2.2. Self-financing strategies

2.2.1. *A portfolio and its trading strategy*

In financial mathematics, a financial market is modeled as a collection of an investor's (or agent's) assets, which is called a portfolio. It may contain stocks, bonds and other securities, bank accounts, investment funds or derivatives (financial instruments that are based on the expected future price movements of an asset). Let S_t^i be the price of the ith asset at time t. All prices S^i are random processes in some probability space $(\Omega, \mathcal{F}, \mathbf{P})$. The collection of all prices $X_t = (S_t^0, S_t^1, \ldots, S_t^N)$, $t \geqslant 0$, is a random vectorial process. Usually, the asset with index 0 is assumed to be riskless (e.g. a bond or bank account) with a constant short-time (or spot) interest rate $r \geqslant 0$, i.e. its price is a non-random process $S_t^0 = S_0^0 e^{rt}$.[1] For short, we call it a bond. The remaining assets $(i = 1, 2, \ldots, N)$ are assumed to be risky, and we call them stocks, denoting their collection by $S_t = (S_t^1, \ldots, S_t^N)$. The investors may change the contents of the portfolio by buying or selling stocks and bonds or their parts. This is modeled by a trading strategy, which is a random vectorial process $\phi_t = (\theta_t^0, \theta_t) = (\theta_t^0, \theta_t^1, \ldots, \theta_t^N)$, $t \geqslant 0$, where θ_t^0 is the number of shares in a bond at time t and θ_t^i $(i = 1, 2, \ldots, N)$ is the number of shares of the ith stock at time t. The portfolio wealth at time t equals

$$V_t = V_t(\phi) = \phi_t \cdot X_t = \theta_t^0 S_t^0 + \theta_t \cdot S_t = \theta_t^0 S_t^0 + \sum_{i=1}^{N} \theta_t^i S_t^i. \qquad [2.1]$$

The first part of the wealth, $\theta_t^0 S_t^0$, is interpreted as the riskless investment, and the second, $\theta_t \cdot S_t$, as a risky investment. Usually, we shall consider trading strategies in a finite time interval $[0, T]$.

EXAMPLE 2.1.– The most simple and well-known example is the Black–Scholes model, where the portfolio consists of two assets, a bond with price $S_t^0 = S_0^0 e^{rt}$ and a stock with price S_t satisfying the linear stochastic differential equation

$$\mathrm{d}S_t = \mu S_t \, \mathrm{d}t + \sigma S_t \, \mathrm{d}B_t$$

---

1 Without loss of generality, we can assume that $S_0^0 = 1$, i.e. $S_t^0 = \mathrm{e}^{rt}$ (it is not important whether we have one \$100 banknote or 100 \$1 banknotes).

or, in the integral form,

$$S_t = S_0 + \mu \int_0^t S_u \, du + \sigma \int_0^t S_u \, dB_u, \quad t \in [0, T]; \qquad [2.2]$$

where the constant μ is called the mean rate of return, and $\sigma > 0$ is called the volatility. This shows, in some sense, the degree of risk of that stock. As we know, the solution of equation [2.2] is the so-called geometric Brownian motion

$$S_t = S_0 \exp\left\{(\mu - \sigma^2/2)t + \sigma B_t\right\}, \quad t \in [0, T]. \qquad [2.3]$$

In the multidimensional Black–Scholes model with N stocks, it is assumed that the stock prices $S^1, \ldots, S^N$ satisfy the SDEs

$$dS_t^i = S_t^i \left(\mu_i \, dt + \sum_{j=1}^d \sigma_{ij} \, dB_t^j\right), \quad i = 1, \ldots, N,$$

where $B^1, \ldots, B^d$ are independent Brownian motions, μ^i are the corresponding mean rates of return and $\sigma = (\sigma_{ij})$ is the matrix of volatilities. In more general models, b^i and σ may be functions of time t or even random processes.

2.2.2. *Self-financing strategies*

In financial mathematics, the main interest is in strategies that are self-financing, which means that the wealth of a portfolio varies without exogenous infusion or withdrawal of money, that is, the purchase of a new asset must be financed by the sale of an old asset. Formally, a self-financing strategy must satisfy the condition

$$dV_t(\phi) = \phi_t \cdot dX_t = \sum_{i=0}^N \theta_t^i \, dS_t^i. \qquad [2.4]$$

Mathematically, it makes sense if we assume that, say, S^i are Itô processes, and θ^i are adapted processes (with respect to some Brownian motion or, more generally, with respect to some filtration $\mathbb{F} = \{\mathcal{F}_t,\ t \in [0,T]\}$). Then condition [2.4] can be rewritten in the integral form

$$V_t(\phi) - V_0(\phi) = \int_0^t \phi_u \cdot \mathrm{d}X_u = \sum_{i=0}^{N} \int_0^t \theta_u^i\, \mathrm{d}S_u^i.$$

Of course, the processes θ^i must satisfy some integrability conditions that guarantee the existence of the integrals on the right-hand side, for example,

$$\int_0^T |\theta_t^0|\, \mathrm{d}t < \infty, \qquad \int_0^T (\theta_t^i)^2\, \mathrm{d}\langle S^i \rangle_t < \infty, \qquad i = 1, \ldots, N.$$

2.2.3. Motivation

This can be motivated as follows. In discrete-time financial models, changes to the contents of a portfolio are allowed only at discrete time moments $t_0, t_1, \ldots, t_k$. In this case, the self-financing condition is written as

$$\phi_{t_k} \cdot X_{t_{k+1}} = \phi_{t_{k+1}} \cdot X_{t_{k+1}}.$$

It can be interpreted as follows. At each time moment t_k, the investor, taking into account known stock prices $S_{t_k}^i$, may redistribute the shares ϕ_{t_k} of assets of the portfolio without any receipt from, or deduction to, an outside source of money. In other words, the wealth of the portfolio at time moment t_{k+1} must remain the same as if the stock prices do not change. Relation with the continuous-time self-financing condition becomes clearer if we rewrite the last equality in the form

$$V_{t_{k+1}}(\phi) - V_{t_k}(\phi) = \phi_{t_{k+1}} \cdot X_{t_{k+1}} - \phi_{t_k} \cdot X_{t_k} = \phi_{t_k} \cdot X_{t_{k+1}} - \phi_{t_k} \cdot X_{t_k}$$
$$= \phi_{t_k} \cdot (X_{t_{k+1}} - X_{t_k}).$$

In view of equation [2.1], the self-financing condition [2.4] can be rewritten as

$$\mathrm{d}V_t(\phi) = \theta_t^0 \mathrm{d}S_t^0 + \theta_t \cdot \mathrm{d}S_t = \theta_t^0 r S_t^0\, \mathrm{d}t + \theta_t \cdot \mathrm{d}S_t$$
$$= \big(V_t(\phi) - \theta_t \cdot S_t\big) r\, \mathrm{d}t + \theta_t \cdot \mathrm{d}S_t$$
$$= r V_t(\phi)\, \mathrm{d}t + \theta_t \cdot (-r S_t\, \mathrm{d}t + \mathrm{d}S_t). \qquad [2.5]$$

Note that in the equation obtained, there is no "riskless" part $\theta^0 S^0$.

2.2.4. *Stock discount*

Denote $\widetilde{S}_t^i = S_t^i/S_t^0 = e^{-rt}S_t^i$, $i = 1, \ldots, N$, and $\widetilde{V}_t(\phi) = V_t(\phi)/S_t^0 = e^{-rt}V_t(\phi)$, where the stock and portfolio prices are discounted with respect to the bond price. Using Itô's formula, by equation [2.5], we have:

$$
\begin{aligned}
d\widetilde{V}_t(\phi) &= -re^{-rt}V_t(\phi)\,dt + e^{-rt}\,dV_t(\phi) \\
&= e^{-rt}\left(-rV_t(\phi)\,dt + dV_t(\phi)\right) \\
&= e^{-rt}\theta_t \cdot \left(-rS_t\,dt + dS_t\right) \\
&= \theta_t \cdot \left(S_t d(e^{-rt}) + e^{-rt}dS_t\right) = \theta_t \cdot d\widetilde{S}_t,
\end{aligned}
$$

or, in the integral form,

$$
\widetilde{V}_t(\phi) = \widetilde{V}_0(\phi) + \int_0^t \theta_u \cdot d\widetilde{S}_u. \qquad [2.6]
$$

The relation obtained shows that every self-financing strategy is completely defined by the initial portfolio wealth $v = V_0(\phi)$ and the risky part of the portfolio, the stock shares $\theta = (\theta^1, \ldots, \theta^N)$, since the bond share θ_t^0 at time t can be calculated from the equalities $V_t(\phi) = S_t^0 \widetilde{V}_t(\phi)$ and $V_t(\phi) = \theta_t^0 S_t^0 + \theta_t \cdot S_t$. Therefore, we obtain the following:

PROPOSITION 2.1.– If $\phi = (\theta^0, \theta)$ is a self-financing strategy, then

$$
V_t = V_t(\phi) = S_t^0\left(v + \int_0^t \theta_u \cdot d\widetilde{S}_u\right), \quad t \geqslant 0, \qquad [2.7]
$$

and

$$
\theta_t^0 = \frac{V_t - \theta_t \cdot S_t}{S_t^0}, \quad t \geqslant 0. \qquad [2.8]
$$

Therefore, it is sometimes convenient, instead of the self-financing strategy $\phi = (\theta^0, \theta)$, to speak of the self-financing strategy θ and to denote the wealth of the portfolio by $V_t(\theta)$, instead of $V_t(\phi) = V_t(\theta^0, \theta)$.

We easily check that the reverse is also true:

PROPOSITION 2.2.– If the portfolio wealth

$$V_t = \theta_t^0 S_t^0 + \theta_t \cdot dS_t$$

satisfies

$$\widetilde{V}_t = V_0 + \int_0^t \theta_u \cdot d\widetilde{S}_u, \quad t \geqslant 0, \tag{2.9}$$

then $\phi = (\theta^0, \theta)$ is a self-financing strategy.

PROOF.– We have:

$$
\begin{aligned}
dV_t &= d(e^{rt}\widetilde{V}_t) = re^{rt}\widetilde{V}_t\, dt + e^{rt}d\widetilde{V}_t \\
&= rV_t\, dt + e^{rt}\theta_t \cdot d\widetilde{S}_t \\
&= rV_t\, dt + e^{rt}\theta_t \cdot d(e^{-rt}S_t) \\
&= rV_t\, dt + e^{rt}\theta_t \cdot (-re^{-rt}S_t\, dt + e^{-rt})\, dS_t \\
&= r(V_t - \theta_t \cdot S_t)\, dt + \theta_t \cdot dS_t \\
&= r\theta_t^0 S_t^0\, dt + \theta_t \cdot dS_t \\
&= \theta_t^0 dS_t^0 + \theta_t \cdot dS_t.\ \square
\end{aligned}
$$

EXAMPLE 2.2.– In the Black–Scholes model, the discounted stock price $\widetilde{S}$ satisfies the equation

$$
\begin{aligned}
d\widetilde{S}_t &= d(e^{-rt}S_t) = -r\,e^{-rt}S_t\, dt + e^{-rt}\, dS_t \\
&= -r\,\widetilde{S}_t\, dt + e^{-rt}S_t(\mu\, dt + \sigma\, dB_t) \\
&= \widetilde{S}_t\big((\mu - r)\, dt + \sigma\, dB_t\big).
\end{aligned}
$$

Therefore, the discounted portfolio wealth under the strategy θ for stock price satisfies the equation

$$d\widetilde{V}_t = \theta_t\, d\widetilde{S}_t = \theta_t\widetilde{S}_t\big((\mu - r)\, dt + \sigma\, dB_t\big).$$

DEFINITION 2.1.– A self-financing strategy ϕ is called an arbitrage strategy if

$$V_0(\phi) = 0, \quad V_T(\phi) \geq 0, \quad \text{and} \quad \mathbf{P}\{V_T(\phi) > 0\} > 0.$$

A financial market is called arbitrage-free or viable[2] if there are no arbitrage strategies.

In an arbitrage-free market, there is no self-financing strategy that allows obtaining a positive profit without any risk.

EXAMPLES 2.1.– 1) For simplicity, we first give an example of arbitrage strategy on an infinite time interval $[0, T] = [0, +\infty]$. Suppose that the financial market consists of a bond without dividends, i.e. $S_t^0 = 1$, $r = 0$, and the stock is $S_t = B_t$. For any $x > 0$, denote $\tau_x = \min\{t \geq 0 : S_t = x\}$. Consider the strategy $\theta_t := \mathbf{1}_{(0,\tau_x]}(t)$. Then, starting at zero wealth $V_0 = 0$, the wealth of the portfolio at time t will be $V_t = \int_0^t \mathbf{1}_{(0,\tau_x]}(u)\,\mathrm{d}B_u = B_{t \wedge \tau_x} \to x$, $t \to \infty$. Thus, $\mathbf{P}\{V_\infty = x > 0\} = 1$, and we have an arbitrage opportunity.

2) It is known that, for every positive random variable $Y \in \mathcal{F}_T$, there exists an adapted random process θ such that $\int_0^T \theta_t \mathrm{d}B_t = Y$.[3] Consider the financial market $(S_t^0, S_t) = (1, B_t)$ and self-financing strategy $\phi_t = (\theta_t^0, \theta_t) = (\int_0^t \theta_s \mathrm{d}B_s - \theta_t B_t, \theta_t)$ (see equations. [2.7]–[2.8]). Then

$$V_0(\phi) = \theta_0^0 \cdot 1 + \theta_0 B_0 = 0,$$

but

$$V_T(\phi) = \int_0^T \theta_s \mathrm{d}B_s = Y > 0.$$

Thus, we again have an opportunity of arbitrage.

In financial mathematics, as a rule, markets with no opportunity of arbitrage are considered. This is not too restrictive, since in real markets, arbitrage opportunities are small and do not last a long time because the market immediately reacts and moves to the equilibrium state. To assure the no-arbitrage, usually, strategies satisfying some integrability or boundedness

---

2 Or we say that there are no arbitrage opportunities.
3 Do not confuse this with the representation $Y = \mathbf{E}Y + \int_0^T H_t \mathrm{d}Bt$ for $Y \in L^2(\Omega)$!

restrictions (to be formulated below) are considered. Such strategies are called *admissible*. Their set must be *sufficiently* wide to have the possibility of calculating and realizing various derivative financial instruments, but at the same time, *not too* wide so that arbitrage strategies could not be admissible.

Thus, we further simply assume that the following (no arbitrage opportunity (**NAO**)) condition is satisfied:

There are no arbitrage strategies in the class of admissible strategies.

THEOREM 2.1.– Let the **NAO** condition be satisfied. Then, any two admissible strategies, having the same portfolio wealth V_T at time moment T, have the same portfolio wealth V_t at every time $t \leqslant T$.

PROOF.– For simplicity, let $t = 0$. Suppose that there are two admissible strategies $\phi = (\theta^0, \theta)$ and $\tilde{\phi} = (\tilde{\theta}^0, \tilde{\theta})$ in a market (S^0, S) such that $V_T(\phi) = V_T(\tilde{\phi})$ but $V_0(\phi) > V_0(\tilde{\phi})$. Consider the new self-financing strategy $\bar{\phi}$, which consists of the strategy $\tilde{\phi} - \phi$ with initial wealth $V_0(\tilde{\phi}) - V_0(\phi) < 0$ for portfolio (S^0, S), together with the investment of the profit $V_0(\phi) - V_0(\tilde{\phi})$ into bond, i.e.

$$\bar{\phi}_t := \tilde{\phi}_t - \phi_t + \left(V_0(\phi) - V_0(\tilde{\phi}), 0\right),$$

and

$$V_t(\bar{\phi}) = (\tilde{\phi}_t - \phi_t) \cdot X_t + \left(V_0(\phi) - V_0(\tilde{\phi})\right)S_t^0.$$

Then, at time moment $t = 0$,

$$V_0(\bar{\phi}) = (\tilde{\phi}_0 - \phi_0) \cdot X_0 + \left(V_0(\phi) - V_0(\tilde{\phi})\right)S_0^0$$
$$= V_0(\tilde{\phi}) - V_0(\phi) + V_0(\phi) - V_0(\tilde{\phi}) = 0,$$

whereas at time moment $t = T$,

$$V_T(\bar{\phi}) = (\tilde{\phi}_T - \phi_T) \cdot X_T + \left(V_0(\phi) - V_0(\tilde{\phi})\right)S_T^0$$
$$= V_T(\tilde{\phi}) - V_T(\phi) + \left(V_0(\phi) - V_0(\tilde{\phi})\right)S_T^0$$
$$= \left(V_0(\phi) - V_0(\tilde{\phi})\right)S_T^0 > 0.$$

Thus, we obtained the arbitrage strategy $\bar{\phi}$, a contradiction to **NAO**. $\square$

PROPOSITION 2.3.– Suppose that the stocks $S = (S^1, S^2)$ in a two-dimensional Black–Scholes model satisfy two stochastic differential equations with the same volatility σ and the same driving Brownian motion B (i.e. affected by the same perturbations),

$$\mathrm{d}S_t^1 = S_t^1(\mu_1\,\mathrm{d}t + \sigma\mathrm{d}B_t) \quad \text{and} \quad \mathrm{d}S_t^2 = S_t^2(\mu_2\,\mathrm{d}t + \sigma\mathrm{d}B_t).$$

Then, in the no-arbitrage market, $\mu_1 = \mu_2$.

PROOF.– Suppose that, say, $\mu_1 > \mu_2$. Consider the strategy where we buy one stock S^1 and sell S_0^1/S_0^2 stocks S^2, i.e. the portfolio wealth is

$$V_t = S_t^1 - \frac{S_0^1}{S_0^2}S_t^2, \quad t \in [0, T].$$

Then, the initial portfolio wealth is

$$V_0 = S_0^1 - \frac{S_0^1}{S_0^2}S_0^2 = 0,$$

whereas the final portfolio wealth is

$$\begin{aligned}
V_T &= S_T^1 - \frac{S_0^1}{S_0^2}S_T^2 \\
&= S_0^1 \exp\left\{\left(\mu_1 - \frac{\sigma^2}{2}\right)T + \sigma B_T\right\} - \frac{S_0^1}{S_0^2}S_0^2 \exp\left\{\left(\mu_2 - \frac{\sigma^2}{2}\right)T + \sigma B_T\right\} \\
&= S_0^1 \exp\left\{\sigma B_T - \frac{\sigma^2}{2}T\right\}\left(e^{\mu_1 T} - e^{\mu_2 T}\right) > 0,
\end{aligned}$$

and so we have the arbitrage strategy, a contradiction. $\square$

REMARK 2.1.– The proposition can be interpreted as follows: In the no-arbitrage market, there cannot be stocks affected by the same perturbations (B and σ) but with different growth tendencies (μ). Also, note that if $\sigma = 0$, i.e. both S^1 and S^2 are riskless instruments, then their investment return must be the same as that of S^0, i.e. $\mu_1 = \mu_2 = r$; otherwise, we could easily construct an arbitrage as in the proposition just proved. We have a slightly more complicated situation when the perturbations are different:

PROPOSITION 2.4.– Suppose that the stocks $S = (S^1, S^2)$ in a two-dimensional Black–Scholes model satisfy two stochastic differential equations with the same driving Brownian motion B:

$$dS_t^1 = S_t^1(\mu_1 \, dt + \sigma_1 dB_t) \quad \text{and} \quad dS_t^2 = S_t^2(\mu_2 \, dt + \sigma_2 dB_t).$$

Then, in the no-arbitrage market,

$$\frac{\mu_1 - r}{\sigma_1} = \frac{\mu_2 - r}{\sigma_2}.$$

PROOF.– Suppose that, say, $\alpha := \frac{\mu_1 - r}{\sigma_1} - \frac{\mu_2 - r}{\sigma_2} > 0$. Consider the self-financing strategy

$$\theta_t = (\theta_t^1, \theta_t^2) := ((\sigma_1 S_t^1)^{-1}, -(\sigma_2 S_t^2)^{-1})$$

and the portfolio with initial value $V_0(\theta) = 0$. Then, by equation [2.7] and example 2.2,

$$V_T(\theta) = e^{rT} \int_0^T \theta_t \cdot d\widetilde{S}_t = e^{rT} \left(\int_0^T \theta_t^1 \, d\widetilde{S}_t^1 + \int_0^T \theta_t^1 \, d\widetilde{S}_t^1 \right)$$

$$= e^{rT} \left(\int_0^T (\sigma_1 S_t^1)^{-1} \widetilde{S}_t^1 ((\mu_1 - r) \, dt + \sigma_1 \, dB_t) \right.$$

$$\left. - \int_0^T (\sigma_2 S_t^2)^{-1} \widetilde{S}_t^2 ((\mu_2 - r) \, dt + \sigma_2 \, dB_t) \right)$$

$$= e^{rT} \int_0^T e^{-rt} \left(\frac{\mu_1 - r}{\sigma_1} - \frac{\mu_2 - r}{\sigma_2} \right) dt = \alpha e^{rT} \int_0^T e^{-rt} \, dt > 0,$$

and so again we have the arbitrage strategy, a contradiction. □

EXERCISE 2.1.– Consider the market with bond price $S_t^0 = S_0^0 e^{rt}$ and stock price S_t, satisfying the equation $dS_t = R_t S_t \, dt$, where R_t is a stochastic process. Prove that, under the no-arbitrage condition, $R_t = r, t \geqslant 0$.

2.3. Option pricing problem: the Black–Scholes model

In the decades to come, the equation will be completely different from what it is now.

Robert Muir-Woods

Recall that in the Black–Scholes model, the portfolio consists of two assets, a riskless asset (bond) S_t^0 and a risky asset (stock) S_t. The bond price is

$$S_t^0 = S_0^0 e^{rt}, \quad t \in [0, T],$$

where $r > 0$ is the constant interest rate. It satisfies the simple ordinary differential equation $dS_t^0 = r S_t^0 \, dt$. The stock price S_t is a random process satisfying the stochastic differential equation $dS_t = \mu S_t \, dt + \sigma S_t dB_t$ or, in the integral form,

$$S_t = S_0 + \mu \int_0^t S_u \, du + \sigma \int_0^t S_u \, dB_u, \quad t \in [0, T]; \qquad [2.10]$$

where the constant μ is called the mean rate of return, and $\sigma > 0$ is the volatility. The solution of the equation is the geometric Brownian motion

$$S_t = S_0 \exp\left\{ (\mu - \sigma^2/2)t + \sigma B_t \right\}, \quad t \in [0, T]. \qquad [2.11]$$

Suppose that we know the initial stock price S_0 and that we want to buy a European call option with maturity T and exercise price K. As we do not know the stock price S_T at time T, a natural question arises: what price we would like or would agree to pay for this option, i.e. what is its true price (premium) at time moment $t = 0$?

Black and Scholes defined this price according to the following principles:

1) The investor after investment, at the initial time moment $t = 0$, has in his portfolio (S_t^0, S_t) this true price $V_0 = \theta_0^0 S_0^0 + \theta_0 S_0$. He can control this portfolio using a self-financing strategy $\theta = (\theta_t^0, \theta_t)$, so that, at maturity T, he would get the same profit $V_T = (S_T - K)^+$ as he would get by buying the option.

2) If the option were offered for any other price, we would have an opportunity of arbitrage.

Note that by theorem 2.1, such a true option price in an arbitrage-free market must be unique.

Thus, let us try to find a self-financing strategy $\tilde{\theta} = (\theta_t^0, \theta_t)$, such that the portfolio wealth $V_t = \theta_t^0 S_t^0 + \theta_t S_t$ satisfies the *terminal* condition

$V_T = (S_T - K)^+$. Note that we may consider the more general problem with a terminal condition of the form $V_T = H := h(S_T)$, where h is a non-negative function. In this context, the random variable H is called a terminal payoff[4] or simply payoff, and a self-financing strategy $\tilde{\theta} = (\theta^0_t, \theta_t)$ such that the terminal wealth $V_T = \theta^0_T S^0_T + \theta_T S_T = H$ is called a hedging strategy leading to H; we also say that a self-financing strategy $\tilde{\theta} = (\theta^0_t, \theta_t)$ hedges (or replicates) the payoff H.

Assume that the portfolio wealth can be written in the form $V_t = F(t, S_t)$, $t \in [0, T]$, with a sufficiently smooth deterministic function $F(t, x)$, $t \in [0, T]$, $x \in \mathbb{R}$.[5] By the self-financing condition, we have:

$$dF(t, S_t) = dV(t) = \theta^0_t \, dS^0_t + \theta_t \, dS_t.$$

On the other hand, by Itô's formula, we have:

$$dF(t, S_t) = F'_t(t, S_t) \, dt + F'_x(t, S_t) \, dS_t + \frac{1}{2} F''_{xx}(t, S_t) \, d\langle S \rangle_t$$

$$= \left(F'_t(t, S_t) + \frac{1}{2}\sigma^2 S^2_t F''_{xx}(t, S_t) \right) dt + F'_x(t, S_t) \, dS_t.$$

From the last two equations, we get that $\theta_t = F'_x(t, S_t)$ and

$$\theta^0_t \, dS^0_t = \left(F'_t(t, S_t) + \frac{1}{2}\sigma^2 S^2_t F''_{xx}(t, S_t) \right) dt.$$

As $dS^0_t = r S^0_t \, dt$, from this we get

$$\theta^0_t S^0_t r \, dt = \left(F'_t(t, S_t) + \frac{1}{2}\sigma^2 S^2_t F''_{xx}(t, S_t) \right) dt,$$

i.e.

$$\theta^0_t S^0_t r = F'_t(t, S_t) + \frac{1}{2}\sigma^2 S^2_t F''_{xx}(t, S_t). \qquad [2.12]$$

---

4 Or contingent claim.
5 Of course, this is a serious restriction. Happily, it is not too strong.

As $F(t, S_t) = \theta_t^0 S_t^0 + \theta_t S_t = \theta_t^0 S_t^0 + F_x'(t, S_t)S_t$, substituting $\theta_t^0 S_t^0 = F(t, S_t) - F_x'(t, S_t)S_t$ into equation [2.12], we get:

$$\left(F(t, S_t) - F_x'(t, S_t)S_t\right)r = F_t'(t, S_t) + \frac{1}{2}\sigma^2 S_t^2 F_{xx}''(t, S_t).$$

Obviously, the latter equality is satisfied, provided that the function F satisfies the partial differential equation

$$r\left(xF_x'(t, x) - F(t, x)\right) + F_t'(t, x) + \frac{1}{2}\sigma^2 x^2 F_{xx}''(t, x) = 0. \qquad [2.13]$$

Note that, in this equation, the terminal condition $V_T = F(T, S_T) = h(S_T)$ (or $F(T, x) = h(x)$) is not reflected in any way. If a function F is its solution, then the random process $V_t = F(t, S_t)$, $t \in [0, T]$ shows the portfolio wealth with some self-financing strategy (θ^0, θ). The latter can be expressed in terms of F:

$$\theta_t^0 = \frac{F(t, S_t) - F_x'(t, S_t)S_t}{S_t^0}, \quad \theta_t = F_x'(t, S_t).$$

Another important property of equation [2.13] is the absence of the coefficient μ. This is important in practice since, in a real financial market, it is difficult to effectively estimate such parameters. Note also that to determine the initial wealth $V_0 = \theta_0^0 S_0^0 + \theta_0 S_0$ of the portfolio, we only need to know the initial values of the portfolio, S_0^0 and S_0.

2.4. The Black–Scholes formula

Finman's Law of Mathematics: Nobody wants to read anyone else's formulas.

A. Bloch, Merfy's Laws

In the Black–Scholes model, to find the European call option price $V_0 = F(0, S_0)$, we have to find the solution $F = F(t, x)$ of equation [2.13], satisfying the terminal condition $F(T, x) = h(x) = (x - K)^+$. Happily, the equation is solvable in an explicit form; without this, it would be difficult to convince practitioners to recognize any model. We shall solve it by using stochastic analysis methods. To this end, we first reverse the time in equation [2.13] by considering the function $u(t, x) := F(T - t, x)$, $t \in [0, T]$, $x > 0$. Then,

$$u'_t(t, x) = -F'_t(T - t, x),$$
$$u'_x(t, x) = F'_x(T - t, x),$$
$$u''_{xx}(t, x) = F''_{xx}(T - t, x),$$

and equation [2.13] becomes

$$u'_t(t, x) = rxu'_x(t, x) + \frac{1}{2}\sigma^2 x^2 u''_{xx}(t, x) - ru(t, x) \qquad [2.14]$$

with the *initial* condition $u(0, x) = h(x)$. For any function $f \in C^2(\mathbb{R})$, we denote

$$Lf(x) := rxf'(x) + \frac{1}{2}\sigma^2 x^2 f''(x).$$

The operator L is the generator of the solution of SDE

$$dX_t = rX_t\, dt + \sigma X_t dB_t.$$

Denote by X_t^x, $t \in [0, T]$, the solution of the latter starting at $X_0^x = x$. Then, by Corollary 1.3(1), the function $v(t, x) := \mathbf{E}h(X_t^x)$ solves the equation $v'_t(t, x) = Lv(t, x)$ with the initial condition $v(0, x) = h(x)$. Then, we directly check that the function $u(t, x) := e^{-rt}v(t, x)$ solves the equation $u'_t(t, x) = Lu(t, x) - ru(t, x)$ (i.e. equation [2.14]) with the same initial condition $u(0, x) = h(x)$.[6]

Now, we find the function $v(t, x) = \mathbf{E}h(X_t^x)$ by applying the explicit formula

$$X_t^x = x\exp\{(r - \sigma^2/2)t + \sigma B_t\}.$$

---

6 The same result can be obtained immediately by applying the Feynman–Kac formula (Corollary 1.3(2)).

Let $\varphi(x) := \frac{1}{\sqrt{2\pi}}e^{-x^2/2}$, $x \in \mathbb{R}$, denote the density of the standard normal distribution, and $\mathcal{N}(x) := \int_{-\infty}^{x} \varphi(y)\,dy$, $x \in \mathbb{R}$, its distribution function. As $B_t \sim \sqrt{t}N(0,1)$, we have:

$$v(t,x) = \int_{\mathbb{R}} h(x\exp\{(r - \sigma^2/2)t + \sigma\sqrt{t}y\})\varphi(y)\,dy. \qquad [2.15]$$

Here, in fact, the function h may be arbitrary. In the case of a call option, we are interested in the concrete function $h(x) = (x - K)^+$, for which we further have:

$$v(t,x) = \int_{\mathbb{R}} (x\exp\{(r - \sigma^2/2)t + \sigma\sqrt{t}y\} - K)^+ \varphi(y)\,dy$$

$$= \int_{D_K} (x\exp\{(r - \sigma^2/2)t + \sigma\sqrt{t}y\} - K)\varphi(y)\,dy,$$

where

$$D_K := \{y \in \mathbb{R} : x\exp\{(r - \sigma^2/2)t + \sigma\sqrt{t}y\} - K \geqslant 0\}$$

$$= \{y \in \mathbb{R} : (r - \sigma^2/2)t + \sigma\sqrt{t}y \geqslant \ln(K/x)\}$$

$$= \left\{y \in \mathbb{R} : y \geqslant \frac{\ln(K/x) - (r - \sigma^2/2)t}{\sigma\sqrt{t}}\right\},$$

i.e. $D_K = [\widetilde{K}, +\infty)$ with $\widetilde{K} = \frac{\ln(K/x) - (r - \sigma^2/2)t}{\sigma\sqrt{t}}$. Thus, continuing, we have:

$$v(t,x) = \int_{\widetilde{K}}^{\infty} (x\exp\{(r - \sigma^2/2)t + \sigma\sqrt{t}y\} - K)\varphi(y)\,dy$$

$$= \frac{1}{\sqrt{2\pi}} \int_{\widetilde{K}}^{\infty} x\exp\{(r - \sigma^2/2)t + \sigma\sqrt{t}y\}e^{-y^2/2}\,dy - K\int_{\widetilde{K}}^{\infty} \varphi(y)\,dy$$

$$= \frac{1}{\sqrt{2\pi}} \int_{\widetilde{K}}^{\infty} x\exp\{-y^2/2 + \sigma\sqrt{t}y + (r - \sigma^2/2)t\}\,dy$$

$$\quad - K(1 - \mathcal{N}(\widetilde{K}))$$

$$= \frac{x}{\sqrt{2\pi}} \int_{\widetilde{K}}^{\infty} \exp\left\{-\frac{y^2 - 2\sigma\sqrt{t}y + \sigma^2 t}{2} + rt\right\}\,dy - K\mathcal{N}(-\widetilde{K})$$

$$= \frac{xe^{rt}}{\sqrt{2\pi}} \int_{\widetilde{K}}^{\infty} e^{-(y-\sigma\sqrt{t})^2)/2} \, dy - K\mathcal{N}(-\widetilde{K})$$

$$= \frac{xe^{rt}}{\sqrt{2\pi}} \int_{\widetilde{K}-\sigma\sqrt{t}}^{\infty} e^{-y^2/2} \, dy - K\mathcal{N}(-\widetilde{K})$$

$$= xe^{rt}\left(1 - \mathcal{N}(\widetilde{K} - \sigma\sqrt{t})\right) - K\mathcal{N}(-\widetilde{K})$$

$$= xe^{rt}\mathcal{N}(\sigma\sqrt{t} - \widetilde{K}) - K\mathcal{N}(-\widetilde{K})$$

$$= xe^{rt}\mathcal{N}(d_1) - K\mathcal{N}(d_2);$$

where we denoted

$$d_1 = d_1(t, x) = \sigma\sqrt{t} - \widetilde{K} = \frac{\ln(x/K) + (r + \sigma^2/2)t}{\sigma\sqrt{t}}$$

and

$$d_2 = d_2(t, x) = -\widetilde{K} = \frac{\ln(x/K) + (r - \sigma^2/2)t}{\sigma\sqrt{t}} = d_1 - \sigma\sqrt{t}.$$

Substituting the expression obtained into the equality

$$F(t, x) = u(T - t, x) = e^{-r(T-t)}v(T - t, x)$$

and denoting, for short, $\tau := T - t$, we finally obtain:

$$F(t, x) = x\mathcal{N}\big(d_1(\tau, x)\big) - Ke^{-r\tau}\mathcal{N}\big(d_2(\tau, x)\big), \tag{2.16}$$

$$d_1(\tau, x) = \frac{\ln(x/K) + (r + \sigma^2/2)\tau}{\sigma\sqrt{\tau}}, \quad d_2(\tau, x) = d_1 - \sigma\sqrt{\tau}. \tag{2.17}$$

To emphasize that the formula obtained gives the *call* option price, the function $F(t, x)$ in equation [2.16] is commonly denoted by $C(t, x)$; moreover, when we want to show its dependence on (some of) the parameters, they are typically denoted by $C(t, x; K, T, r, \sigma)$ or, say $C(t, x; K)$.

Let us summarize the results obtained above:

The price of the European call option in the Black–Scholes model is

$$V_0 = C(0, S_0) = S_0 \mathcal{N}\big(d_1(T, S_0)\big) - K e^{-rT} \mathcal{N}\big(d_2(T, S_0)\big), \qquad [2.18]$$

where d_1 and d_2 are the functions defined in equation [2.17].

The corresponding wealth process is

$$V_t = C(t, S_t) = S_t \mathcal{N}\big(d_1(T - t, S_t)\big) - K e^{-r(T-t)} \mathcal{N}\big(d_2(T - t, S_t)\big),$$
$$t \in [0, T],$$
$$[2.19]$$

realized by the self-financing strategy (θ_t^0, θ_t), $t \in [0, T]$, with

$$\theta_t = C'_x(t, S_t) \quad \text{and} \quad \theta_t^0 = \frac{C(t, S_t) - \theta_t S_t}{S_t^0}. \qquad [2.20]$$

Formula [2.18] is the famous Black–Scholes formula for call option price. Note that the option price does not depend on the mean rate of return μ (usually unknown) but does depend on the volatility σ.

The European put option price $P(t, x)$ in the Black–Scholes model is calculated similarly, replacing the end condition $h(x) = (x - K)^+$ by the end condition $h(x) = (K - x)^+$:

$$P(t, x) = K e^{-r(T-t)} \mathcal{N}\big(-d_2(T - t, x)\big) - x \mathcal{N}\big(-d_1(T - t, x)\big). \quad [2.21]$$

The call and put option prices satisfy the following relation, which is called the *call–put parity*:

$$C(t, x) - P(t, x) = x - K e^{-r(T-t)}, \qquad [2.22]$$

which is convenient for calculating one of the option prices (call or put) when the other is known. Indeed, using the equality $\mathcal{N}(x) + \mathcal{N}(-x) = 1$, we have

$$C(t, x) - P(t, x) = x \mathcal{N}(d_1) - K e^{-r\tau} \mathcal{N}(d_2)$$
$$- K e^{-r(T-t)} \mathcal{N}(-d_2) - x \mathcal{N}(-d_1)$$

$$= x\big(\mathcal{N}(d_1) + \mathcal{N}(-d_1)\big)$$
$$\quad - K\mathrm{e}^{-r(T-t)}\big(\mathcal{N}(d_2) + \mathcal{N}(-d_2)\big)$$
$$= x - K\mathrm{e}^{-r(T-t)}.$$

Typical graphs of prices of call and put options are shown in Figures 2.1 and 2.2.

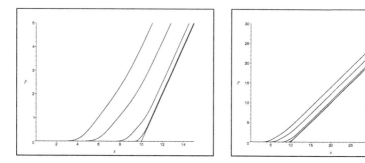

Figure 2.1. *Call option prices $C(t,x)$ as functions of x at different scales; $K = 10$, $T = 10$, $r = 0.05$, $\sigma = 0.1$, $t = 0, 5, 9, 9.9, 10$*

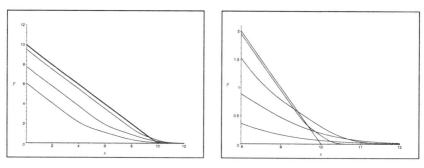

Figure 2.2. *Put option prices $P(t,x)$ as functions of x at different scales; $K = 10$, $T = 10$, $r = 0.05$, $\sigma = 0.1$, $t = 0, 5, 9, 9.9, 10$*

In Figure 2.3, we illustrate the hedging strategies for a call option on a simulated trajectory of the stock price S_t in two cases where $S_T > K$ and $S_T < K$. We see that, in both cases, the riskless investment is negative, i.e. we are constantly borrowing money.

This is not a coincidence. In the next subsection, we will show that the hedging strategy, or the so-called Delta, of the call option

$$\theta_t = C'_x(t, S_t) = \mathcal{N}(d_1(T - t, S_t)) \in (0, 1),$$

and thus the risky investment $\theta_t S_t$ is always positive. This implies that, on the contrary, the riskless investment in the hedging portfolio is always negative:

$$\theta_t^0 S_t^0 = V_t - \theta_t S_t = -Ke^{-r(T-t)}\mathcal{N}(d_2(T - t, S_t)) < 0.$$

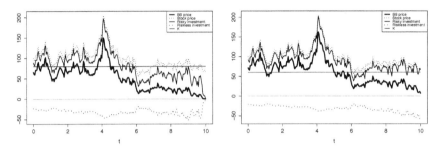

Figure 2.3. *Hedging of call option.* $T = 10$, $r = 0.05$, $\sigma = 0.4$, $S_0 = 100$.
Left: $S_T < K = 80$; *right:* $S_T > K = 60$

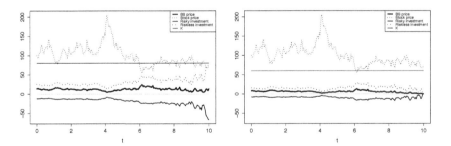

Figure 2.4. *Hedging of put option.* $T = 10$, $r = 0.05$, $\sigma = 0.4$, $S_0 = 100$.
Left: $S_T < K = 80$; *right:* $S_T > K = 60$

The situation with the put option is opposite (see Figure 2.4, which illustrates the hedging of a put option on the same simulated trajectory of stock price):

$$\theta_t = P'_x(t, S_t) = -\mathcal{N}(-d_1(T - t, S_t)) \in (-1, 0),$$

and

$$\theta_t^0 S_t^0 = V_t - \theta_t S_t = K \mathrm{e}^{-r(T-t)} \mathcal{N}\big(- d_2(T - t, S_t)\big) > 0,$$

which means that we constantly invest on the riskless bond.

2.4.1. *Option Greeks*[7]

The Greeks are the quantities representing the sensitivity of the price of options to a change in underlying parameters. They are defined as the derivatives of the option price with respect to the corresponding parameters. The most popular Greeks are the following:

Delta $\quad \Delta := \dfrac{\partial C}{\partial x};$

Vega $\quad \nu := \dfrac{\partial C}{\partial \sigma};$

Theta $\quad \Theta := \dfrac{\partial C}{\partial t};$

Rho $\quad \rho := \dfrac{\partial C}{\partial r};$

Gamma $\quad \Gamma := \dfrac{\partial^2 C}{\partial x^2}.$

Note that the self-financing strategy for the call option can be written in the form $\theta_t = \Delta(t, S_t)$, $t \in [0, T]$. The Greeks of put options are denoted by the same letters, with price C replaced with P.

PROPOSITION 2.5.– Shortly denoting $d_i = d_i(\tau, x) = d_i(T - t, x)$, $i = 1, 2,$ we have the following formulas:

$$\Delta = \frac{\partial C}{\partial x} = \mathcal{N}(d_1); \qquad\qquad \frac{\partial P}{\partial x} = -\mathcal{N}(-d_1); \quad [2.23]$$

$$\nu = \frac{\partial C}{\partial \sigma} = x\sqrt{\tau}\varphi(d_1) = \mathrm{e}^{-r\tau} K \sqrt{\tau}\varphi(d_2); \qquad \frac{\partial P}{\partial \sigma} = \frac{\partial C}{\partial \sigma}; \qquad [2.24]$$

---

7 The name is used because these parameters of options are denoted by Greek letters. They are also called risk sensitivities, risk measures, or hedge parameters.

$$\Theta = \frac{\partial C}{\partial t} = -rK\mathrm{e}^{-r\tau}\mathcal{N}(d_2) - \frac{x\sigma\varphi(d_1)}{2\sqrt{\tau}};$$

$$\frac{\partial P}{\partial t} = rK\mathrm{e}^{-r\tau}\mathcal{N}(-d_2) - \frac{x\sigma\varphi(d_1)}{2\sqrt{\tau}};$$ [2.25]

$$\rho = \frac{\partial C}{\partial r} = \tau K\mathrm{e}^{-r\tau}\mathcal{N}(d_2); \qquad\qquad \frac{\partial P}{\partial r} = -\tau K\mathrm{e}^{-r\tau}\mathcal{N}(-d_2);$$ [2.26]

$$\Gamma = \frac{\partial^2 C}{\partial x^2} = \frac{\varphi(d_1)}{x\sigma\sqrt{\tau}} \qquad\qquad \frac{\partial^2 P}{\partial x^2} = \frac{\partial^2 C}{\partial x^2}.$$ [2.27]

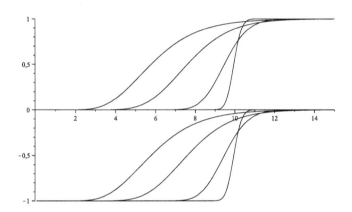

Figure 2.5. *Delta* $(\Delta(t,x))$ *for call (above) and put (below) options;* $x \in [0,15]$, $K = 10$, $T = 10$, $r = 0.05$, $\sigma = 0.1$. $t = 0; 5; 9; 9.9; 10$

PROOF.– Let us first check the formulas for the call option. We will need the following identity:

$$x\varphi(d_1) - K\mathrm{e}^{-r\tau}\varphi(d_2) = 0.$$ [2.28]

To this end, note that

$$x\varphi(d_1) = K\mathrm{e}^{-r\tau}\varphi(d_2) \iff \frac{x}{K}\mathrm{e}^{r\tau} = \frac{\varphi(d_2)}{\varphi(d_1)}$$

$$\iff \ln(x/K) + r\tau = \frac{d_1^2 - d_2^2}{2},$$

where the last equality can be checked directly:

$$\frac{d_1^2 - d_2^2}{2} = \frac{1}{2}(d_1 + d_2)(d_1 - d_2) = \frac{1}{2}(2d_1 - \sigma\sqrt{\tau})\sigma\sqrt{\tau}$$

$$= \ln(x/K) + (r + \sigma^2/2)\tau - \frac{1}{2}\sigma^2\tau$$

$$= \ln(x/K) + r\tau.$$

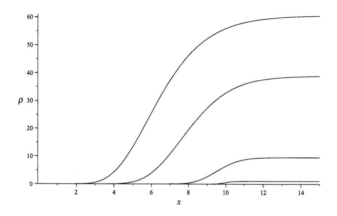

Figure 2.6. *Rho* $(\nu(t,x))$ *for call option;* $x \in [0, 15]$, $K = 10$, $T = 10$, $r = 0.05$, $\sigma = 0.1$, $t = 0; 5; 9; 9.9; 10$

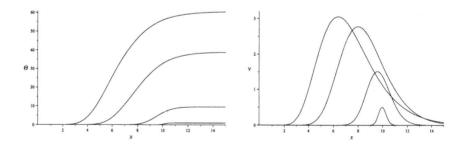

Figure 2.7. *Theta* $(\Theta(t,x))$ *and Vega* $(\nu(t,x))$ *for call option;* $x \in [0, 15]$, $K = 10$, $T = 10$, $r = 0.05$, $\sigma = 0.1$, $t = 0; 5; 9; 9.9; 10$

Delta

$$\Delta = \frac{\partial C}{\partial x} = \mathcal{N}(d_1) + x\varphi(d_1)\frac{\partial d_1}{\partial x} - K\mathrm{e}^{-r\tau}\varphi(d_2)\frac{\partial d_2}{\partial x}$$

$$= \mathcal{N}(d_1) + \frac{1}{x\sigma\sqrt{\tau}}\left(x\varphi(d_1) - K\mathrm{e}^{-r\tau}\varphi(d_2)\right) \overset{[2.28]}{=} \mathcal{N}(d_1);$$

Vega

$$\nu = \frac{\partial C}{\partial \sigma} = x\varphi(d_1)\frac{\partial d_1}{\partial \sigma} - K\mathrm{e}^{-r\tau}\varphi(d_2)\frac{\partial d_2}{\partial \sigma}$$

$$= x\varphi(d_1)\frac{\partial d_1}{\partial \sigma} - K\mathrm{e}^{-r\tau}\varphi(d_2)\left(\frac{\partial d_1}{\partial \sigma} - \sqrt{\tau}\right)$$

$$= x\varphi(d_1)\left(\frac{\partial d_2}{\partial \sigma} + \sqrt{\tau}\right) - K\mathrm{e}^{-r\tau}\varphi(d_2)\frac{\partial d_2}{\partial \sigma}$$

$$= x\varphi(d_1)\sqrt{\tau} + \frac{\partial d_2}{\partial \sigma}\left(x\varphi(d_1) - K\mathrm{e}^{-r\tau}\varphi(d_2)\right)$$

$$\overset{[2.28]}{=} x\sqrt{\tau}\varphi(d_1);$$

Theta

$$\Theta = \frac{\partial C}{\partial t} = x\varphi(d_1)\frac{\partial d_1}{\partial t} - rK\mathrm{e}^{-r\tau}\mathcal{N}(d_2) - K\mathrm{e}^{-r\tau}\varphi(d_2)\frac{\partial d_2}{\partial t}$$

$$= -rK\mathrm{e}^{-r\tau}\mathcal{N}(d_2) + \frac{\partial d_1}{\partial t}x\varphi(d_1) - K\mathrm{e}^{-r\tau}\varphi(d_2)\left(\frac{\partial d_1}{\partial t} + \frac{\sigma}{2\sqrt{\tau}}\right)$$

$$= -rK\mathrm{e}^{-r\tau}\mathcal{N}(d_2) - \frac{\sigma K\mathrm{e}^{-r\tau}\varphi(d_2)}{2\sqrt{\tau}} + \frac{\partial d_1}{\partial t}\underbrace{\left(x\varphi(d_1) - K\mathrm{e}^{-r\tau}\varphi(d_2)\right)}_{0}$$

$$= -rK\mathrm{e}^{-r\tau}\mathcal{N}(d_2) - \frac{\sigma x\mathrm{e}^{-r\tau}\varphi(d_1)}{2\sqrt{\tau}};$$

Rho

$$\rho = \frac{\partial C}{\partial r} = x\varphi(d_1)\frac{\partial d_1}{\partial r} + \tau K\mathrm{e}^{-r\tau}\mathcal{N}(d_2) - K\mathrm{e}^{-r\tau}\varphi(d_2)\frac{\partial d_2}{\partial r}$$

$$= -\tau K\mathrm{e}^{-r\tau}\mathcal{N}(d_2) + \frac{\partial d_1}{\partial r}x\varphi(d_1) - K\mathrm{e}^{-r\tau}\varphi(d_2)\frac{\partial d_2}{\partial t}$$

$$= -\tau K e^{-r\tau} \mathcal{N}(d_2) + \frac{\partial d_1}{\partial r} \big(\underbrace{x \varphi(d_1) - K e^{-r\tau} \varphi(d_2)}_{0} \big)$$

$$= -\tau K e^{-r\tau} \mathcal{N}(d_2);$$

Gamma $\Gamma = \dfrac{\partial^2 C}{\partial x^2} = \dfrac{\partial \Delta}{\partial x} = \varphi(d_1) \dfrac{\partial d_1}{\partial x} = \dfrac{\varphi(d_1)}{x \sigma \sqrt{\tau}}.$

The Greeks for put options are obtained by using the Greeks for call options and the following equalities, which are obtained by differentiation of the parity identity [2.22] with respect to the corresponding parameter:

$$\frac{\partial P}{\partial x} = \frac{\partial C}{\partial x} - 1, \quad \frac{\partial P}{\partial \sigma} = \frac{\partial C}{\partial \sigma}, \quad \frac{\partial P}{\partial t} = \frac{\partial C}{\partial t} + r K e^{-r\tau},$$

$$\frac{\partial P}{\partial r} = \frac{\partial C}{\partial r} - \tau K e^{-r\tau}, \quad \frac{\partial^2 P}{\partial x^2} = \frac{\partial^2 C}{\partial x^2}. \qquad \qquad \Box$$

EXERCISE 2.2.– Check that the function $F(t, x) = x$ satisfies the Black–Scholes PDF [2.13]. What is the corresponding payoff in the Black–Scholes model? Calculate the price of the corresponding option.

EXERCISE 2.3.– In the Black–Scholes model, consider the options with payoffs

$$H_1 = S_T 1_{\{S_T > K\}} \quad \text{(asset-or-nothing option)}$$

and

$$H_2 = 1_{\{S_T > K\}} \quad \text{(cash-or-nothing or digital option)}.$$

Find the prices of these options (without repeating the calculations).

EXERCISE 2.4.– In the Black–Scholes model, consider the options with payoffs

$$H_3 = |S_T - K| \quad \text{(straddle option with basis price } K)$$

and

$$H_4 = \begin{cases} K_1 - S_T & \text{if } S_T < K_1, \\ 0 & \text{if } K_1 \leqslant S_T \leqslant K_2, \\ S_T - K_2 & \text{if } S_T > K_2 \end{cases}$$

(straddle option with basis prices $K_1 < K_2$).

Find the prices of these options (without repeating the calculations).

EXERCISE 2.5.– *Gap option*. The payoff of a gap call option is given by

$$H = (S_T - G)\mathbf{1}_{\{S_T \geqslant K\}}$$

with $G, K \geqslant 0$, where, in general, $G \neq K$. Find the price of this option.

Hint: Write the payoff of the gap call option in the following form:

$$H = (S_T - K)^+ - (G - K)\mathbf{1}_{\{S_T \geqslant K\}}$$

and apply the Black–Scholes formula and the answer for H_2 in exercise 2.3.

EXERCISE 2.6.– Show the call option price $C(t, x; K, T, r, \sigma)$ as a strictly increasing function with respect to σ with limits

$$\lim_{\sigma \downarrow 0} C(t, x; K, T, r, \sigma) = \left(x - Ke^{-r(T-t)}\right)^+ \quad \text{and}$$

$$\lim_{\sigma \to +\infty} C(t, x; K, T, r, \sigma) = x.$$

Give a financial interpretation of these properties.

EXERCISE 2.7.– Show the following symmetry properties of the call option price:

$$C(t, x; K, T, r, \sigma) = C(0, x; K, T - t, r, \sigma)$$
$$C(t, ax; aK, T, r, \sigma) = aC(t, x; K, T, r, \sigma) \quad \text{for all } a > 0.$$

EXERCISE 2.8.– Consider the Black–Scholes model with varying deterministic volatility, where the stock price follows the equation

$$dS_t = \mu St\, dt + \sigma(t)S_t dB_t, \quad S_0 = x,$$

where σ is a continuous function. Show (without repeating the calculations) that the call option price can be written in the following form:

$$C(0, x) = \widetilde{\mathbf{E}}(S_T - K)^+ = x\mathcal{N}(D_1) - Ke^{-rT}\mathcal{N}(D_2)$$

and find D_1 and D_2.

EXERCISE 2.9.– In the Black–Scholes model, we consider the option with payoff $H = S_T^n$. Show that the option price $V_t = F(t, S_t)$, $t \in [0, T]$, is of the following form:

$$F(t, x) = f(t, T)x^n. \qquad [2.29]$$

Find the function $f(t, T)$.

EXERCISE 2.10.– Check directly that the function

$$u(t, x) := e^{-rt} \int_{\mathbb{R}} h(x \exp\{(r - \sigma^2/2)t + \sigma\sqrt{t}y\})\varphi(y)\, dy,$$

$$x > 0,\ t > 0,$$

satisfies equation [2.14] with initial condition $u(0, x) := \lim_{t\downarrow 0} u(t, x) = h(x)$, i.e. the function $F(t, x) = u(T - t, x)$ satisfies the Black–Scholes PDE [2.13] with terminal condition $F(T, x) = h(x)$.

REMARK 1.1.– Assume that differentiation under the integral sign is justified.

EXERCISE 2.11.– Show that the Greek Delta

$$\Delta \to 1 \text{ as } t \uparrow T \text{ if } x > K \text{ ("in-the-money" call option)}$$

and

$$\Delta \to 0 \text{ as } t \uparrow T \text{ if } x < K \text{ ("out–of-the-money" call option)}.$$

2.5. Risk-neutral probabilities: alternative derivation of the Black–Scholes formula

2.5.1. *Risk-neutral probability*

Recall that, in the Black–Scholes model, the stock price satisfies the equation

$$dS_t = \mu S_t\, dt + \sigma S_t\, dB_t,$$

and the discounted stock price $\widetilde{S}_t = e^{-rt}S_t$ satisfies the equation

$$d\widetilde{S}_t = \widetilde{S}_t((\mu - r)\, dt + \sigma\, dB_t).$$

Denote $\lambda = (\mu - r)/\sigma$. The number λ is called *market price of risk*.[8] Then, the last equation can be written as

$$d\widetilde{S}_t = \sigma\widetilde{S}_t(\lambda\,dt + dB_t)$$

Denoting $\widetilde{B}_t = B_t + \lambda t$, $t \in [0, T]$ (Brownian motion with drift), we get the equation

$$d\widetilde{S}_t = \sigma\widetilde{S}_t\,d\widetilde{B}_t$$

By the Girsanov theorem (theorem 1.24), there exists a probability $\widetilde{\mathbf{P}} \sim \mathbf{P}$ such that the Brownian motion with a drift, $\widetilde{B}_t := B_t + \lambda t$, $t \in [0, T]$, is a Brownian motion with respect to $\widetilde{\mathbf{P}}$ (i.e. in the probability space $(\Omega, \mathcal{F}, \widetilde{\mathbf{P}})$). From the equation, we see that the discounted stock price $\widetilde{S}$ is a martingale with respect to $\widetilde{\mathbf{P}}$ (or a $\widetilde{\mathbf{P}}$-martingale for short). Such a probability $\widetilde{\mathbf{P}}$ is called a risk-neutral probability or a martingale probability.

In proposition 2.4 (p. 40), we saw that, in the two-dimensional Black–Scholes model, where the stock values satisfy the equations

$$dS_t^1 = S_t^1(\mu_1\,dt + \sigma_1 dB_t) \quad \text{and} \quad dS_t^2 = S_t^2(\mu_2\,dt + \sigma_2 dB_t),$$

there is no arbitrage iff

$$\lambda := \frac{\mu_1 - r}{\sigma_1} = \frac{\mu_2 - r}{\sigma_2},$$

i.e. the market prices of the risk of S^1 and S^2 coincide. In such a case, again denoting $\widetilde{B}_t = B_t + \lambda t$, $t \in [0, T]$, we now obtain the equations

$$d\widetilde{S}_t^i = \sigma_i\widetilde{S}_t^i\,d\widetilde{B}_t, \quad i = 1, 2.$$

Taking the probability $\widetilde{\mathbf{P}} \sim \mathbf{P}$ such that $\widetilde{B}_t = B_t + \lambda t$, $t \in [0, T]$, is a $\widetilde{\mathbf{P}}$-Brownian motion, we see that both discounted stock prices $\widetilde{S}^1$ and $\widetilde{S}^2$ are $\widetilde{\mathbf{P}}$-martingales. In the general case, we have the following definition.

---

8 Interpretation: in a financial market, the return of a risky investment must be, on average, higher than that of a riskless investment. The market price of risk is the relative (with respect to risk σ) rate of extra return above the risk-free rate r.

DEFINITION 2.2.– A probability $\widetilde{\mathbf{P}}$ is called a risk-neutral probability or a martingale probability if it is equivalent to the probability $\mathbf{P}$, and all discounted stock prices $\widetilde{S}_t^i = \mathrm{e}^{-rt} S_t^i$, $t \in [0, T]$, are $\widetilde{\mathbf{P}}$-martingales.

Recall that two probabilities $\widetilde{\mathbf{P}}$ and $\mathbf{P}$ are said to be equivalent if, for all events A,

$$\widetilde{P}(A) > 0 \iff P(A) > 0.$$

From the definition, it is clear why $\widetilde{\mathbf{P}}$ is called a *martingale* probability. The term *risk-neutral* probability is used because the average growth rates of all (risky) stocks $S_t^i = \widetilde{S}_t^i \, \mathrm{e}^{rt}$ with respect to $\widetilde{\mathbf{P}}$ are equal to the interest rate r of the (riskless) bond, i.e. $\widetilde{\mathbf{P}}$ is "neutral" with respect to the risk.

Therefore, we have shown the existence of a risk-neutral probability in the one- and two-dimensional Black–Scholes models. The existence of such a probability in a discrete-time model is equivalent to the no-arbitrage opportunity (**NAO**) condition. Moreover, in the discrete-time case, all self-financing strategies appear to be admissible. An analogous general fact in continuous-time models is known as *the first fundamental theorem of financial mathematics*; however, its formulation is much more subtle. To avoid this, we therefore replace the **NAO** condition by the close condition of the existence of risk-neutral probability:

(**RNP**) *In the market, there exists a risk-neutral probability.*

So, every arbitrage-free financial market likely exists in two parallel spaces, the initial "risky" probability space $(\Omega, \mathcal{F}, \mathbf{P})$, where the "true" portfolio wealth process is defined, and, at the same time, the "riskless" probability space $(\Omega, \mathcal{F}, \widetilde{\mathbf{P}})$, where the average growth rates of stocks are equal to the riskless interest rate r. The first space serves for modeling price dynamics, whereas the second, as we will see, serves as a tool for option pricing.

As the wealth of a self-financing portfolio satisfies equation [2.6], i.e.

$$\widetilde{V}_t(\phi) = \widetilde{V}_0(\phi) + \int_0^t \theta_u \cdot \mathrm{d}\widetilde{S}_u,$$

we see that the discounted wealth $\widetilde{V}_t(\phi)$ is an integral with respect to the $\widetilde{\mathbf{P}}$-martingale $\widetilde{S}$. Therefore, in "good" cases, it is also a $\widetilde{\mathbf{P}}$-martingale.

DEFINITION 2.3.– A self-financing strategy ϕ is called $\widetilde{\mathbf{P}}$-admissible if the discounted portfolio wealth $\widetilde{V}_t(\phi)$ is a $\widetilde{\mathbf{P}}$-martingale.

REMARK 2.2.– In the Black–Scholes model, the discounted portfolio wealth $\widetilde{V}_t = \widetilde{V}_t(\phi)$ satisfies the equation

$$\mathrm{d}\widetilde{V}_t = \sigma\theta_t\widetilde{S}_t\,\mathrm{d}\widetilde{B}_t.$$

A sufficient condition for $\widetilde{V}$ to be a $\widetilde{\mathbf{P}}$-martingale is $\theta\widetilde{S} \in H^2[0,T]$, i.e.

$$\widetilde{\mathbf{E}}\left[\int_0^T (\theta_t\widetilde{S}_t)^2\,\mathrm{d}t\right] < +\infty, \qquad\qquad [2.30]$$

where $\widetilde{\mathbf{E}}$ denotes the expectation with respect to $\widetilde{\mathbf{P}}$.

PROPOSITION 2.6.– There are no arbitrage strategies in the class of $\widetilde{\mathbf{P}}$-admissible strategies.

PROOF.– Suppose, on the contrary, that a $\widetilde{\mathbf{P}}$-admissible strategy ϕ is an arbitrage strategy. Then, $\widetilde{V}_0(\phi) = V_0(\phi) = 0$, $\widetilde{V}_T(\phi) = V_T(\phi)\mathrm{e}^{-rT} \geqslant 0$, and $\mathbf{P}\{\widetilde{V}_T(\phi) > 0\} > 0$. As $\widetilde{\mathbf{P}} \sim \mathbf{P}$, we also have $\widetilde{\mathbf{P}}\{\widetilde{V}_T(\phi) > 0\} > 0$. Therefore, $\widetilde{\mathbf{E}}[\widetilde{V}_T(\phi)] > 0$. On the other hand, since $\widetilde{V}(\phi)$ is a $\widetilde{\mathbf{P}}$-martingale, we have $\widetilde{\mathbf{E}}[\widetilde{V}_T(\phi)] = \widetilde{\mathbf{E}}[\widetilde{V}_0(\phi)] = 0$, a contradiction. $\square$

REMARK 2.3.– Another common way to avoid the opportunity of arbitrage when condition **RNP** is satisfied is the limitation to the strategies ϕ for which the process $M_t = \int_0^t \phi_s\mathrm{d}X_s$ is bounded from below (i.e. admissible strategies cannot lead to bankruptcy). In this case,

$$\widetilde{V}_t = V_0 + \int_0^t \theta_u \cdot \mathrm{d}\widetilde{S}_u, \quad t \in [0,T],$$

is bounded from below local $\widetilde{\mathbf{P}}$-martingale. Therefore, $\widetilde{\mathbf{E}}\widetilde{V}_t \leqslant \widetilde{\mathbf{E}}V_0$, $t \in [0,T]$. Thus, if $V_0 = 0$, then $\widetilde{\mathbf{E}}\widetilde{V}_T \leqslant 0$, and, as in proposition 2.6, we see that the arbitrage is impossible. To show this, we have to recall the notion of a *local* martingale. A process M_t, $t \geqslant 0$, is called a local martingale if there exists an increasing sequence of Markov moments $\{\tau_n\}$ such that $\lim_{n\to\infty}\tau_n = +\infty$ and, for all $n \in \mathbb{N}$, the processes $M_t^n := M(t \wedge \tau_n)$, $t \geqslant 0$, are all martingales. A typical example of a local martingale is a stochastic integral $M_t = \int_0^t X_s\,\mathrm{d}B_s$ – it suffices to take

$\tau_n = \min\{t \geqslant 0 : |M_t| = n\}$. If a local martingale M is bounded from below, then by the Fatou lemma,[9] we have

$$\mathbf{E}M_t \leqslant \liminf_{n\to\infty} \mathbf{E}M_t^n = \liminf_{n\to\infty} \mathbf{E}M_0^n = \mathbf{E}M_0.$$

DEFINITION 2.4.– A payoff (or contingent claim) is any non-negative ($\mathcal{F}_T$-measurable) random variable H. A self-financing strategy ϕ is said to replicate (or hedge) a payoff H if $V_T(\phi) = H$.

The initial portfolio wealth $V_0 = V_0(\phi)$ under such a strategy is called a fair price of H. In the arbitrage-free market, any other price would produce the possibility of arbitrage (see theorem 2.1).

DEFINITION 2.5.– A financial market is said to be complete if every payoff is replicatable by an admissible self-financing strategy.

THEOREM 2.2 (Second fundamental theorem of FM).– The arbitrage-free market is complete if and only if the risk-neutral probability is unique.

We further consider a complete arbitrage-free market with risk-neutral probability $\widetilde{\mathbf{P}}$.

THEOREM 2.3.– If a payoff H has a finite expectation $\widetilde{\mathbf{E}}H$, then the portfolio wealth of the strategy replicating the payoff equals

$$V_t = V_t(\phi) = \widetilde{\mathbf{E}}\left[e^{-r(T-t)}H\big|\mathcal{F}_t\right]. \qquad [2.31]$$

The fair price of the payoff is $V_0 = \widetilde{\mathbf{E}}[e^{-rT}H]$.

PROOF.– Every discounted portfolio wealth $\widetilde{V}_t(\phi) = e^{-rt}V_t(\phi)$ with $\widetilde{\mathbf{P}}$-admissible strategy ϕ is a $\widetilde{\mathbf{P}}$-martingale. If a self-financing strategy ϕ replicates the payoff H, then $V_T(\phi) = H$. From this, we obtain

$$e^{-rt}V_t(\phi) = \widetilde{\mathbf{E}}\left[e^{-rT}V_T(\phi)\big|\mathcal{F}_t\right] = \widetilde{\mathbf{E}}\left[e^{-rT}H\big|\mathcal{F}_t\right],$$

and, therefore,

$$V_t(\phi) = \widetilde{\mathbf{E}}\left[e^{-r(T-t)}H\big|\mathcal{F}_t\right].\square$$

---

[9] If $X = \lim_{n\to\infty} X_n$ a.s. and there is a r.v. Y such that $\mathbf{E}|Y| < \infty$ and $X_n \geqslant Y$, $n \in \mathbb{N}$, then $\mathbf{E}X \leqslant \liminf_{n\to\infty} \mathbf{E}X_n$.

So, we know the fair price of a payoff. How can we realize it and find the corresponding replicating self-financing strategy $\phi = (\theta^0, \theta)$, for which $V_T(\phi) = H$? By theorem 2.31 and self-financing condition [2.6], we have

$$\widetilde{\mathbf{E}}[e^{-rT}H|\mathcal{F}_t] = V_t(\phi)e^{-rt} = \widetilde{V}_t(\phi) = V_0 + \int_0^t \theta_u \, d\widetilde{S}_u.$$

This is an integral equation with a stochastic integral of the unknown process θ on the right-hand side. For example, in the case of the Black–Scholes model, we have the equation

$$\widetilde{\mathbf{E}}[e^{-rT}H|\mathcal{F}_t] = V_0 + \sigma \int_0^t \theta_u \widetilde{S}_u \, d\widetilde{B}_u.$$

How can we solve such an equation? If the integral were ordinary, we could find θ by simple differentiation with respect to t. This does not work in the case of a stochastic integral. In the case of diffusion models[10] with European payoffs $H = h(S_T)$, the equation can be solved by means of Itô's formula. We illustrate this for the Black–Scholes model. In this case,

$$V_t = V_t(\phi) = \widetilde{\mathbf{E}}\left[e^{-r(T-t)}h(S_T)\big|\mathcal{F}_t\right] \tag{2.32}$$

with stock price S satisfying the equation

$$dS_t = rS_t \, dt + \sigma S_t \, d\widetilde{B}_t,$$

where $\widetilde{B}$ is a Brownian motion with respect to the risk-neutral probability $\widetilde{\mathbf{P}}$. Indeed, recalling that $\widetilde{B}_t = B_t + \lambda t$ with $\lambda = (\mu - r)/\sigma$, we have

$$dS_t = \mu S_t \, dt + \sigma S_t \, dB_t = \mu S_t \, dt + \sigma S_t \, d(\widetilde{B}_t - \lambda t)$$

$$= \mu S_t \, dt + \sigma S_t \, d\left(\widetilde{B}_t - \frac{\mu - r}{\sigma}t\right)$$

$$= rS_t \, dt + \sigma S_t \, d\widetilde{B}_t.$$

To calculate the conditional expectation with respect to $\mathcal{F}_t$ or, equivalently, with respect to S_t (because of the Markov property of S), we write S_T as a

---

10 When the stock prices are diffusion processes.

function of S_t and a random variable independent of S_t:

$$S_T = S_0 \exp\left\{\sigma \widetilde{B}_T + (r - \sigma^2/2)T\right\}$$

$$= S_0 \exp\left\{\sigma \widetilde{B}_t + (r - \sigma^2/2)t\right\} \cdot \exp\left\{\sigma(\widetilde{B}_T - \widetilde{B}_t) + (r - \sigma^2/2)(T - t)\right\}$$

$$= S_t \exp\left\{\sigma(\widetilde{B}_T - \widetilde{B}_t) + (r - \sigma^2/2)(T - t)\right\}. \qquad [2.33]$$

As $\widetilde{B}_T - \widetilde{B}_t \perp\!\!\!\perp S_t \in \mathcal{F}_t$ and $\widetilde{B}_T - \widetilde{B}_t \stackrel{d}{=} \xi\sqrt{T-t}$ with $\xi \sim N(0,1)$, taking the conditional expectation with respect to $\mathcal{F}_t$, by the Markov property of S and formula [1.4], we get

$$\widetilde{\mathbf{E}}\left[h(S_T)\big|\mathcal{F}_t\right] = \widetilde{\mathbf{E}}\left[h(S_T)\big|S_t\right]$$

$$= \mathbf{E}\,h\left(x\exp\left\{\sigma\xi\sqrt{T-t} + (r - \sigma^2/2)(T-t)\right\}\right)\Big|_{x=S_t}$$

$$= v(T - t, S_t)$$

with

$$v(t, x) = \mathbf{E}\,h\left(x\exp\left\{\sigma\xi\sqrt{t} + (r - \sigma^2/2)t\right\}\right)$$

$$= \int_{\mathbb{R}} h\left(x\exp\left\{\sigma y\sqrt{t} + (r - \sigma^2/2)t\right\}\right)\varphi(y)\,\mathrm{d}y,$$

where, as before, $\varphi(y) = \frac{1}{\sqrt{2\pi}}e^{-y^2/2}$, $y \in \mathbb{R}$, is the density of the standard normal distribution.

Finally, we obtain:

$$V_t = V_t(\phi) = \widetilde{\mathbf{E}}\left[e^{-r(T-t)}h(S_T)\big|\mathcal{F}_t\right] = e^{-r(T-t)}v(T - t, S_t).$$

Denoting $F(t, x) = e^{-r(T-t)}v(T - t, x)$, we obtain the portfolio wealth under the strategy ϕ replicating the payoff $h(S_T)$:

$$V_t(\phi) = F(t, S_t),$$

which coincides with that obtained in sections 2.3 and 2.4. Therefore, in the same way, we derive the Black–Scholes formula [2.18] for call option (and [2.21] for put option), and applying the equalities

$$V_t(\phi) = F(t, S_t) = \theta_t^0 S_t^0 + \theta_t S_t \quad \text{and} \quad \mathrm{d}V_t(\phi) = \theta_t^0 \mathrm{d}S_t^0 + \theta_t \mathrm{d}S_t,$$

we find the replicating strategy $\phi = (\theta^0, \theta)$ defined by formulas [2.20].

Finally, it is worth noting the advantages of this new derivation of the Black–Scholes formula:

– we did not derive and need not solve the Black–Scholes PDE equation for the function F;

– we have found the wealth of replicating strategy by using the existence of a risk-neutral probability and formula [2.31] for the portfolio wealth under the replicating self-financing strategy;

– this approach allows us to expect success in finding the prices of options with more complicate payoffs H.

EXERCISE 2.12.– Prove the call–put parity relation (equation [2.22]) by using formula [2.31].

EXERCISE 2.13.– Consider the Black–Scholes model with payoff $H = h(S_T)$, where $h \in C^2(\mathbb{R})$. Prove the following "synthesized" formula for the portfolio wealth under the replicating self-financing strategy:

$$V_t = \int_0^\infty h''(K) C(t, S_t; K) \, \mathrm{d}K, \quad 0 \leqslant t \leqslant T.$$

Hint: First, show the relation $h(x) = \int_0^\infty h''(K)(x - K)^+ \, \mathrm{d}K$ and then combine it with formula [2.31] to $H = h(S_T)$.

EXERCISE 2.14.– Find the function $f(t, T)$ in exercise 2.4.2.9 by applying formula [2.31].

EXERCISE 2.15.– Show the inequalities

$$(x - Ke^{-r(T-t)})^+ \leqslant C(t, x; K, T) \leqslant x, \quad x \geqslant 0.$$

EXERCISE 2.16.– Suppose that the bond price S^0 in the Black–Scholes model is also stochastic and is a geometric Brownian motion satisfying the equation $dS_t^0 = S_t^0(r \, dt + \rho \, dB_t)$.

i) Check that the discounted stock price $\widetilde{S}_t = S_t/S_t^0$ satisfies the SDE

$$d\widetilde{S}_t = \widetilde{S}_t\left[(\mu - r + \rho^2 - \rho\sigma)\,dt + (\sigma - \rho)\,dB_t\right].$$

ii) Show that there exist a probability $\widetilde{\mathbf{P}} \sim \mathbf{P}$ and a Brownian motion $\widetilde{B}$ with respect to $\widetilde{\mathbf{P}}$ such that

$$d\widetilde{S}_t = (\sigma - \rho)\widetilde{S}_t\,d\widetilde{B}_t.$$

iii) Let $H = h(S_T)$ be a payoff at maturity T. Prove that its fair price at time t is $V_t = S_t\widetilde{\mathbf{E}}(S_T^{-1}H)$; in particular, $V_0 = \widetilde{\mathbf{E}}(S_T^{-1}H)$.

EXERCISE 2.17.– Consider the so-called Bachelier model $S_t = S_0 + \mu t + \sigma B_t$ with $r = 0$.

i) Find a formula for the call option price.

ii) Derive a Black–Scholes-type PDE for this model.

EXERCISE 2.18.– Consider the model given by the SDE $dS_t = \mu S_t dt + \sigma dB_t$.

i) Solve the SDE.

ii) Find $\lambda \in \mathbb{R}$ such that $\widetilde{B}_t = B_t + \lambda t$, $t \in [0, T]$, is a Brownian motion with respect to the risk-neutral probability.

iii) Find the portfolio wealth V_t, $t \in [0, T]$, of the strategy replicating the payoff $H = \exp\{S_T\}$.

EXERCISE 2.19.– Consider the multidimensional Black–Scholes model with bond $S_t^0 = e^{rt}$ and N stocks $S^1, \ldots, S^N$ satisfying the SDEs

$$dS_t^i = S_t^i\left(\mu_i\,dt + \sum_{j=1}^{d}\sigma_{ij}\,dB_t^j\right), \quad i = 1, \ldots, N,$$

where $B^1, \ldots, B^d$ are independent Brownian motions, $\mu_i \in \mathbb{R}$ are the corresponding mean rates of return, and $\sigma = (\sigma_{ij})$ is the matrix of volatilities. Suppose that the linear equation system

$$\sum_{j=1}^{d}\sigma_{ij}\lambda_j = \mu_i - r, \quad i = 1, \ldots, N,$$

has a solution $\lambda = (\lambda_1, \ldots, \lambda_N) \in \mathbb{R}^N$ (this is the case, e.g. where $N = d$ and the matrix σ is invertible; also cf. proposition 2.4.). Let a probability $\widetilde{\mathbf{P}} \sim \mathbf{P}$ be such that the Brownian motions with drifts $\widetilde{B}_t^j = B_t^j + \lambda_j t$, $t \in [0, T]$, $j = 1, \ldots, N$, are independent Brownian motions with respect to $\widetilde{\mathbf{P}}$. (Such an equivalent probability exists by a multidimensional version of the Girsanov theorem.)

Show that $\widetilde{\mathbf{P}}$ is a risk-neutral probability.

EXERCISE 2.20.– Consider the multidimensional Black–Scholes model as in exercise 2.5.2.19 with $N = d$ and invertible volatility matrix $\sigma = (\sigma_{ij})$. Find the price of the call option on the first stock, i.e. when the payoff is $(S_T^1 - K)^+$.

2.6. American options in the Black–Scholes model

Contrary to European options, which do not give the right to exercise the option before the maturity T, an American option may be exercised at any time before maturity T.

Naturally, at the time moment $t \leqslant T$, the holder of the option can make the decision to exercise the option based only on the information available until moment t. Therefore, the exercise moment is a non-negative random variable τ that does not "know" the future. Formally, the latter property is defined as the requirement that $\{\tau \leqslant t\} \in \mathcal{F}_t$ for all $t \in [0, T]$. Such random variables are called stopping times (or Markov moments).

The payoff in the case of an American option is a non-negative random process h_t, $t \in [0, T]$. For simplicity, we shall suppose that $h_t = h(S_t)$ with non-negative function $h \in C[0, \infty)$ of at most linear growth: $\exists k, b \geqslant 0$: $h(x) \leqslant kx + b$, $x \geqslant 0$. In the "standard" cases of call and put options, $h(x) = (x - K)^+$ and $h(x) = (K - x)^+$, respectively.

DEFINITION 2.6.– We say that a $\widetilde{\mathbf{P}}$-admissible strategy $\phi = (\theta^0, \theta)$ replicates (or hedges) a payoff h if $V_t(\phi) \geqslant h(S_t)$, $t \in [0, T]$.

THEOREM 2.4.– Denote

$$u(t, x) = \sup_{\tau \in \mathcal{T}_{t,T}} \widetilde{\mathbf{E}} \left[e^{-r(\tau - t)} h \left(x \exp\{\sigma(\widetilde{B}_\tau - \widetilde{B}_t) + (r - \sigma^2/2)(\tau - t)\} \right) \right],$$

where $\mathcal{T}_{t,T}$ is the class of all stopping times taking values in the interval $[t, T]$. There exists a strategy $\bar{\phi}$ such that $V_t(\bar{\phi}) = u(t, S_t)$, $t \in [0, T]$. Moreover, $V_t(\phi) \geqslant u(t, S_t)$, $t \in [0, T]$, for any other strategy ϕ.

PROPOSITION 2.7.– If $h(x) = (x - K)^+$ in the theorem, then

$$u(t, x) = F(t, x), \quad x \in \mathbb{R},$$

where $F(t, x)$ is the function from the BS formula. In other words, the price of the American call option coincides with that of the European call option.

PROOF.– For simplicity, we consider $t = 0$. It suffices to show that, for every topping time $\tau \in \mathcal{T}_{0,T}$,

$$\widetilde{\mathbf{E}}\left(e^{-r\tau}(S_\tau - K)^+\right) \leqslant \widetilde{\mathbf{E}}\left(e^{-rT}(S_T - K)^+\right)$$
$$= \widetilde{\mathbf{E}}\left((\widetilde{S}_T - e^{-rT}K)^+\right).$$

On the other hand, since $\widetilde{S}$ is a $\widetilde{\mathbf{P}}$-martingale, we have

$$\widetilde{\mathbf{E}}\left((\widetilde{S}_T - e^{-rT}K)^+\big|\mathcal{F}_\tau\right) \geqslant \widetilde{\mathbf{E}}\left((\widetilde{S}_T - e^{-rT}K)\big|\mathcal{F}_\tau\right) = \widetilde{S}_\tau - e^{-rT}K$$
$$\geqslant \widetilde{S}_\tau - e^{-r\tau}K.$$

As the left-hand side is non-negative, we have

$$\widetilde{\mathbf{E}}\left((\widetilde{S}_T - e^{-rT}K)^+\big|\mathcal{F}_\tau\right) \geqslant (\widetilde{S}_\tau - e^{-r\tau}K)^+. \qquad \square$$

2.7. Exotic options

Having finished school, you will soon forget 70% of what you have learned, but you will never forget that you have overcome all that.
Bronislovas Burgis

The options considered thus far are often called vanilla options. More complicated options that depend not only on the end stock prices, but also on the dynamics of stock prices on the whole time interval $[0, T]$, are called exotic options.

2.7.1. *Barrier options*

These are rather popular, as their prices are less than those of vanilla options, but their valuation is more complicated. Their payoffs at the maturity T depend on whether stock prices in the time interval $[0, T]$ have been reached a certain, a priori prescribed, level (barrier). Let us first consider call options.

– *Down-and-out call options* (DOC): The option holder loses his right of exercising it if the stock price S in the interval $[0, T]$ falls below the prescribed level L (we suppose that $L < S_0$, since, otherwise, this price would obviously be 0). In the opposite case, he gets the payoff $h(S_T)$ with $h(x) = (x - K)^+$. In other words, the option price at the initial moment $t = 0$ equals

$$DOC(S_0, K, L) = \widetilde{\mathbf{E}} \left[e^{-rT} (S_T - K)^+ \mathbf{1}_{\{\min_{t \leqslant T} S_t > L\}} \right].$$

Let us denote τ_L the moment at which the underlying stock price reaches the barrier L for the first time:

$$\tau_L = \min\{t : S_t \leqslant L\} = \min\{t : S_t = L\}.$$

Then, the DOC option prices can be written as

$$DOC(S_0, K, L) = \widetilde{\mathbf{E}} \left[e^{-rT} (S_T - K)^+ \mathbf{1}_{\{\tau_L > T\}} \right].$$

– *Down-and-in call options* (DIC): The option holder gets his payoff only in the case where the stock price S in the interval $[0, T]$ reaches the level L before time T. The corresponding option price at time moment $t = 0$ equals

$$DIC(S_0, K, L) = \widetilde{\mathbf{E}} \left[e^{-rT} (S_T - K)^+ \mathbf{1}_{\{\min_{t \leqslant T} S_t \leqslant L\}} \right]$$

$$= \widetilde{\mathbf{E}} \left[e^{-rT} (S_T - K)^+ \mathbf{1}_{\{\tau_L \leqslant T\}} \right].$$

If $L \geqslant S_0$, the DIC option price coincides with the European call option price $C(S_0, K) = \widetilde{\mathbf{E}}[e^{-rT}(S_T - K)^+]$. Also note that

$$DOC(S_0, K, L) + DIC(S_0, K, L) = C(S_0, K). \tag{2.34}$$

– *Up-and-out* and *Up-and-in call options* (UOC and UIC): The options are defined similarly to DOC and DIC options, with the barrier H reached from below (we suppose that $H > S_0$). The UOC option price at the initial time moment $t = 0$ equals

$$UOC(S_0, K, H) = \widetilde{\mathbf{E}} \left[e^{-rT}(S_T - K)^+ \mathbf{1}_{\{\max_{t \leqslant T} S_t < H\}} \right]$$

$$= \widetilde{\mathbf{E}} \left[e^{-rT}(S_T - K)^+ \mathbf{1}_{\{\tau_H > T\}} \right],$$

where $\tau_H = \min\{t : S_t \geqslant H\} = \min\{t : S_t = H\}$. The UIC option price at time moment $t = 0$ is

$$UIC(S_0, K, H) = \widetilde{\mathbf{E}} \left[e^{-rT}(S_T - K)^+ \mathbf{1}_{\{\max_{t \leqslant T} S_t \geqslant H\}} \right]$$

$$= \widetilde{\mathbf{E}} \left[e^{-rT}(S_T - K)^+ \mathbf{1}_{\{\tau_H \leqslant T\}} \right].$$

As in the DOC and DIC cases, we have

$$UOC(S_0, K, H) + UIC(S_0, K, H) = C(S_0, K).$$

– *Barrier* **put** *options* DOP, DIP, UOP and UIP are defined in exactly the same way, by replacing the payoff function by $h(x) = (K - x)^+$.

2.7.2. *Other exotic options*

– *Lookback options.* They are similar to call and put options, but their exercise price equals the maximum and minimum of the stock price in the interval $[0, T]$. For example, the lookback call and put option payoffs are $S_T - \min_{t \leqslant T} S_t$ and $\max_{t \leqslant T} S_t - S_T$, respectively.

– *Asian options.* The payoffs depend on the average stock price in the interval $[0, T]$. Their payoffs equal

$$\left(\frac{1}{T} \int_0^T S_t \, dt - K \right)^+$$

for call option and

$$\left(K - \frac{1}{T} \int_0^T S_t \, dt \right)^+$$

for put option.

– *Asset-or-nothing option.* The payoff $S_T \mathbf{1}_{\{S_T \geqslant K\}}$.

– *Digital option.* The payoff $\mathbf{1}_{\{S_T > K\}}$ ("cash-or-nothing").

2.7.3. *Barrier option pricing*

As an example, let us find the prices of the DIC option

$$DIC(S_0, L, K) = e^{-rT}\, \widetilde{\mathbf{E}}\left[(S_T - K)^+ \mathbf{1}_{\{\tau_L \leqslant T\}}\right]. \qquad [2.35]$$

We have

$$S_t = S_0 \exp\left(\sigma \widetilde{B}_t + (r - \sigma^2/2)t\right) = S_0 \exp\left(\sigma \widehat{B}_t\right)$$

with $\widehat{B}_t = \widetilde{B}_t + \mu t$, $\mu = \sigma^{-1}(r - \sigma^2/2)$. Let us express the moment τ_L of reaching the barrier L as a function of S and $\widehat{B}$:

$$\tau_L = \inf\{t : S_t \leqslant L\} = \inf\{t : \sigma \widehat{B}_t \leqslant \ln \frac{L}{S_0}\} = \inf\{t : \widehat{B}_t \leqslant l\} =: \widehat{\tau}_l,$$

where we denoted $l = \frac{1}{\sigma} \ln \frac{L}{S_0} < 0$.

There are two possible ways of calculating the expressions of the form $\widetilde{\mathbf{E}}(f(S_T)\mathbf{1}_{\{\tau_L \leqslant T\}})$: by using the Girsanov theorem, which allows us, by the change of probability, to make the process $\widehat{B}$ to be a Brownian motion, or by using the formula of the joint density of the pair $(\widehat{B}_t, \widehat{m}_t)$, where $\widehat{B}$ is a Brownian motion with drift μ, and $\widehat{m}$ is its minimum process $\widehat{m}_t = \min_{s \leqslant t} \widehat{B}_s$. The first way seems to be slightly simpler.

Thus, define yet one probability $\widehat{\mathbf{P}}$ whose density with respect to $\widetilde{\mathbf{P}}$ is

$$\frac{d\widehat{\mathbf{P}}}{d\widetilde{\mathbf{P}}} = \exp\left(-\mu \widetilde{B}_T - \frac{1}{2}\mu^2 T\right),$$

so that $\widehat{B}$ is a Brownian motion with respect to $\widehat{\mathbf{P}}$. Then,

$$\frac{d\widetilde{\mathbf{P}}}{d\widehat{\mathbf{P}}} = \exp\left(\mu \widetilde{B}_T + \frac{1}{2}\mu^2 T\right) = \exp\left(\mu(\widehat{B}_T - \mu T) + \frac{1}{2}\mu^2 T\right)$$

$$= \exp\left(\mu \widehat{B}_T - \frac{1}{2}\mu^2 T\right).$$

From this, we get the equality

$$\widetilde{\mathbf{E}}\{f(S_T)\mathbf{1}_{\{\tau_L \leqslant T\}}\} = \widehat{\mathbf{E}}\{f(S_0 e^{\sigma \widehat{B}_T}) \exp\left(\mu \widehat{B}_T - \frac{\mu^2 T}{2}\right) \mathbf{1}_{\{\widehat{\tau}_l \leqslant T\}}\}.$$

As $\widehat{B}$ is a Brownian motion with respect to $\widehat{\mathbf{P}}$, we can take off the hats⌒on the right-hand side. Thus, denoting $S_0 = x$, the price formula [2.35] for the DIC option can be expressed in the following form:

$$e^{rT} DIC(x, L, K) = \widetilde{\mathbf{E}}\left[(S_T - K)^+ \mathbf{1}_{\{\tau_L \leqslant T\}}\right]$$

$$= \mathbf{E}\left\{(xe^{\sigma B_T} - K)^+ \exp\left(\mu B_T - \frac{\mu^2 T}{2}\right)\mathbf{1}_{\{\tau_l \leqslant T\}}\right\},$$

$$\text{with} \quad \tau_l = \inf\{t : B_t \leqslant l\}. \tag{2.36}$$

Rewrite the claim $(xe^{\sigma B_T} - K)^+$ as the difference

$$(xe^{\sigma B_T} - K)^+ = (xe^{\sigma B_T} - K)\mathbf{1}_{\{xe^{\sigma B_T} - K \geqslant 0\}}$$

$$= xe^{\sigma B_T}\mathbf{1}_{\{B_T \geqslant k\}} - K\mathbf{1}_{\{B_T \geqslant k\}}, \quad k = \frac{1}{\sigma}\ln\frac{K}{x}.$$

Substituting this expression into formula [2.36], we obtain

$$e^{rT} DIC(x, L, K) = \exp\left(-\frac{\mu^2 T}{2}\right)\left[x\,\mathbf{E}\left(e^{(\sigma+\mu)B_T}\mathbf{1}_{\{B_T \geqslant k\}}\mathbf{1}_{\{\tau_l \leqslant T\}}\right)\right.$$

$$\left. - K\mathbf{E}\left(e^{\mu B_T}\mathbf{1}_{\{B_T \geqslant k\}}\mathbf{1}_{\{\tau_l \leqslant T\}}\right)\right]$$

$$= \exp\left(-\frac{\mu^2 T}{2}\right)\left[x\Psi(\sigma + \mu) - K\Psi(\mu)\right]$$

$$\text{with} \quad \Psi(z) := \mathbf{E}\left(e^{zB_T}\mathbf{1}_{\{B_T \geqslant k\}}\mathbf{1}_{\{\tau_l \leqslant T\}}\right).$$

Note that $\{\tau_l \leqslant T\} = \{m_T := \min_{s \leqslant T} B_s \leqslant l\}$, and therefore,

$$\Psi(z) := \mathbf{E}\left(e^{zB_T}\mathbf{1}_{\{B_T \geqslant k\}}\mathbf{1}_{\{m_T \leqslant l\}}\right). \tag{2.37}$$

Now it is time to recall the joint distribution of (B_T, m_T) (equation [1.1]):

$$\mathbf{P}\{B_T \geqslant x, m_T \geqslant y\} = \begin{cases} \mathcal{N}\left(\frac{-x}{\sqrt{T}}\right) - \mathcal{N}\left(\frac{2y-x}{\sqrt{T}}\right), & y \leqslant 0, \ x \geqslant y, \\ \mathcal{N}\left(\frac{-y}{\sqrt{T}}\right) - \mathcal{N}\left(\frac{y}{\sqrt{T}}\right), & y \leqslant 0, \ x \leqslant y, \\ 0, & y \geqslant 0. \end{cases}$$

Therefore,

$$\mathbf{P}\{B_T \geqslant x,\, m_T \leqslant y\} = \mathbf{P}\{B_T \geqslant x\} - \mathbf{P}\{B_T \geqslant x,\, m_T \geqslant y\}$$

$$= \begin{cases} \mathcal{N}\left(\frac{2y-x}{\sqrt{T}}\right), & y \leqslant 0,\, x \geqslant y, \\ 1 - \mathcal{N}\left(\frac{x}{\sqrt{T}}\right) - \mathcal{N}\left(\frac{-y}{\sqrt{T}}\right) + \mathcal{N}\left(\frac{y}{\sqrt{T}}\right), & y \leqslant 0,\, x \leqslant y, \\ 1 - \mathcal{N}\left(\frac{x}{\sqrt{T}}\right), & y \geqslant 0. \end{cases}$$

Differentiating with respect to x, we get the density of B_T on the event $\{m_T \leqslant y\}$:

$$p(x; T, y) = \frac{\mathbf{P}\{B_T \in \mathrm{d}x,\, m_T \leqslant y\}}{\mathrm{d}x} = \begin{cases} \frac{1}{\sqrt{T}}\varphi\left(\frac{2y-x}{\sqrt{T}}\right), & y \leqslant 0,\, x \geqslant y, \\ \frac{1}{\sqrt{T}}\varphi\left(\frac{x}{\sqrt{T}}\right), & y \leqslant 0,\, x \leqslant y, \\ \frac{1}{\sqrt{T}}\varphi\left(\frac{x}{\sqrt{T}}\right), & y \geqslant 0, \end{cases}$$

where, as usual, $\varphi(x) = \frac{1}{2\pi}\mathrm{e}^{-x^2/2}$ is the standard normal density.

If $k \leqslant l < 0$ (i.e. $K \leqslant L$), then, substituting this expression into equation [2.37], we get

$$\Psi(z) = \int_{-\infty}^{\infty} \mathrm{e}^{zx} \mathbf{1}_{\{x \geqslant k\}} p(x; T, l)\, \mathrm{d}x = \int_{k}^{\infty} \mathrm{e}^{zx} p(x; T, l)\, \mathrm{d}x$$

$$= \left(\int_{k}^{l} + \int_{l}^{\infty} \right) \mathrm{e}^{zx} p(x; T, l)\, \mathrm{d}x$$

$$= \int_{k}^{l} \mathrm{e}^{zx} \frac{1}{\sqrt{T}} \varphi\left(\frac{x}{\sqrt{T}}\right) \mathrm{d}x + \int_{l}^{\infty} \mathrm{e}^{zx} \frac{1}{\sqrt{T}} \varphi\left(\frac{2l-x}{\sqrt{T}}\right) \mathrm{d}x$$

$$= \frac{1}{\sqrt{2\pi T}} \left[\int_{k}^{l} \exp\left\{ zx - \frac{x^2}{2T} \right\} \mathrm{d}x + \int_{l}^{\infty} \exp\left\{ zx - \frac{(2l-x)^2}{2T} \right\} \mathrm{d}x \right]$$

$$= \frac{1}{\sqrt{2\pi T}} \left[\int_{k}^{l} \exp\left\{ \frac{Tz^2}{2} - \frac{(x - zT)^2}{2T} \right\} \mathrm{d}x \right.$$

$$\left. + \int_{l}^{\infty} \exp\left\{ \frac{Tz^2}{2} + 2lz - \frac{(x - (zT + 2l))^2}{2T} \right\} \mathrm{d}x \right]$$

$$= \exp\left\{\frac{Tz^2}{2}\right\}\left[\mathcal{N}\left(\frac{l-zT}{\sqrt{T}}\right) - \mathcal{N}\left(\frac{k-zT}{\sqrt{T}}\right)\right]$$

$$+ \exp\left\{\frac{Tz^2}{2} + 2lz\right\}\left[1 - \mathcal{N}\left(-\frac{l+zT}{\sqrt{T}}\right)\right]$$

$$= \exp\left\{\frac{Tz^2}{2}\right\}\left[\mathcal{N}(z_1) - \mathcal{N}(z_2) + \exp\{2lz\}\mathcal{N}(z_3)\right],$$

where

$$z_1 = \frac{1}{\sqrt{T}}(l - zT), \quad z_2 = \frac{1}{\sqrt{T}}(k - zT), \quad z_3 = \frac{1}{\sqrt{T}}(l + zT).$$

Summarizing and recalling that $\mu = \sigma^{-1}(r - \sigma^2/2)$, $\sigma + \mu = \sigma^{-1}(r + \sigma^2/2)$, $l = -\sigma^{-1}\ln(x/L)$, and $k = -\sigma^{-1}\ln(x/K)$, we get

$$DIC(x, L, K) = \exp\left(-\frac{\mu^2 T}{2} - rT\right)[x\Psi(\sigma + \mu) - K\Psi(\mu)]$$

$$= x\exp\left(-\frac{\mu^2 T}{2} - rT + \frac{T(\sigma + \mu)^2}{2}\right)$$

$$\times\left[\mathcal{N}\left(\frac{1}{\sqrt{T}}(l - (\sigma + \mu)T) - \mathcal{N}\left(\frac{1}{\sqrt{T}}(k - (\sigma + \mu)T\right)\right.$$

$$\left. + \exp\{2l(\sigma + \mu)\}\mathcal{N}\left(\frac{1}{\sqrt{T}}(l + (\sigma + \mu)T)\right)\right]$$

$$- K\exp\left(-\frac{\mu^2 T}{2} - rT + \frac{T\mu^2}{2}\right)$$

$$\times\left[\mathcal{N}\left(\frac{1}{\sqrt{T}}(l - \mu T)\right) - \mathcal{N}\left(\frac{1}{\sqrt{T}}(k - \mu T)\right)\right.$$

$$\left. + \exp\{2l\mu\}\mathcal{N}\left(\frac{1}{\sqrt{T}}(l + \mu T)\right)\right]$$

$$= x\left[\mathcal{N}\left(-\frac{(r + \sigma^2/2)T + \ln(x/L)}{\sigma\sqrt{T}}\right) - \mathcal{N}\left(-\frac{(r + \sigma^2/2)T + \ln(x/K)}{\sigma\sqrt{T}}\right)\right.$$

$$\left. + \left(\frac{L}{x}\right)^{\frac{2r}{\sigma^2}+1}\mathcal{N}\left(\frac{(r + \sigma^2/2)T - \ln(x/L)}{\sigma\sqrt{T}}\right)\right]$$

$$- K \exp(-rT) \left[\mathcal{N}\left(-\frac{(r - \sigma^2/2)T + \ln(x/L)}{\sigma\sqrt{T}} \right) \right.$$

$$- \mathcal{N}\left(-\frac{(r - \sigma^2/2)T + \ln(x/K)}{\sigma\sqrt{T}} \right)$$

$$\left. + \left(\frac{L}{x}\right)^{\frac{2r}{\sigma^2}-1} \mathcal{N}\left(\frac{(r - \sigma^2/2)T - \ln(x/L)}{\sigma\sqrt{T}} \right) \right]$$

$$= x \left[\mathcal{N}(z_1) - \mathcal{N}(z_2) + \left(\frac{L}{x}\right)^{\frac{2r}{\sigma^2}+1} \mathcal{N}(z_3) \right]$$

$$- K\mathrm{e}^{-rT} \left[\mathcal{N}(z_4) - \mathcal{N}(z_5) + \left(\frac{L}{x}\right)^{\frac{2r}{\sigma^2}-1} \mathcal{N}(z_6) \right],$$

where

$$z_1 = \frac{1}{\sigma\sqrt{T}} \left[(r + \sigma^2/2)T + \ln(x/K) \right], \quad z_4 = z_1 - \sigma\sqrt{T}, \qquad \text{[2.38a]}$$

$$z_2 = \frac{1}{\sigma\sqrt{T}} \left[(r + \sigma^2/2)T + \ln(x/L) \right], \quad z_5 = z_2 - \sigma\sqrt{T}, \qquad \text{[2.38b]}$$

$$z_3 = \frac{1}{\sigma\sqrt{T}} \left[(r + \sigma^2/2)T - \ln(x/L) \right], \quad z_6 = z_3 - \sigma\sqrt{T}. \qquad \text{[2.38c]}$$

In the case $k \geqslant l$ (i.e. $K \geqslant L$), we similarly get

$$\Psi(z) = \int_{-\infty}^{\infty} \mathrm{e}^{zx} \mathbf{1}_{\{x \geqslant k\}} p(x; T, l)\, \mathrm{d}x = \int_{k}^{\infty} \mathrm{e}^{zx} p(x; T, l)\, \mathrm{d}x$$

$$= \int_{k}^{\infty} \mathrm{e}^{zx} \frac{1}{\sqrt{T}} \varphi\left(\frac{2l - x}{\sqrt{T}}\right) \mathrm{d}x$$

$$= \frac{1}{\sqrt{2\pi T}} \int_{k}^{\infty} \exp\left\{ zx - \frac{(2l - x)^2}{2T} \right\} \mathrm{d}x$$

$$= \frac{1}{\sqrt{2\pi T}} \int_{k}^{\infty} \exp\left\{ \frac{Tz^2}{2} + 2lz - \frac{(x - (zT + 2l))^2}{2T} \right\} \mathrm{d}x$$

$$= \exp\left\{ \frac{Tz^2}{2} + 2lz \right\} \left[1 - \mathcal{N}\left(\frac{k - (zT + 2l)}{\sqrt{T}} \right) \right]$$

$$= \exp\left\{\frac{Tz^2}{2} + 2lz\right\}\mathcal{N}\left(\frac{2l - k + zT}{\sqrt{T}}\right)$$

$$= \exp\left\{\frac{Tz^2}{2} + 2lz\right\}\mathcal{N}(z_7)$$

and

$$DIC(x, L, K) = x\left(\frac{L}{x}\right)^{\frac{2r}{\sigma^2}+1}\mathcal{N}(z_7) - Ke^{-rT}\left(\frac{L}{x}\right)^{\frac{2r}{\sigma^2}-1}\mathcal{N}(z_8),$$

where

$$z_7 = \frac{1}{\sigma\sqrt{T}}\left[(r + \sigma^2/2)T + \ln(L^2/xK)\right], \quad z_8 = z_7 - \sigma\sqrt{T}.$$

Joining the two formulas, we finally have:

$$DIC(x, K, L) = \begin{cases} x\left[\mathcal{N}(z_1) - \mathcal{N}(z_2) + \left(\frac{L}{x}\right)^{\frac{2r}{\sigma^2}+1}\mathcal{N}(z_3)\right] \\ \quad - Ke^{-rT}\left[\mathcal{N}(z_4) - \mathcal{N}(z_5) + \left(\frac{L}{x}\right)^{\frac{2r}{\sigma^2}-1}\mathcal{N}(z_6)\right], \\ \qquad\qquad\qquad\qquad x > L > K, \\ x\left(\frac{L}{x}\right)^{\frac{2r}{\sigma^2}+1}\mathcal{N}(z_7) - Ke^{-rT}\left(\frac{L}{x}\right)^{\frac{2r}{\sigma^2}-1}\mathcal{N}(z_8), \\ \qquad\qquad\qquad\qquad x > L, \ K > L, \\ C(x, K), \qquad\qquad\qquad x \leqslant L, \end{cases} \quad [2.39]$$

with z_i, $i = 1, \ldots, 8$, defined in (2.38a–d).

The formula obtained can be expressed by means of the European call option price $C(x, K)$. Recall that

$$C(x, K) = x\mathcal{N}(d_1) - Ke^{-rT}\mathcal{N}(d_2),$$

where

$$d_1 = d_1(x, K) = \frac{\ln(x/K) + (r + \sigma^2/2)T}{\sigma\sqrt{T}}, \quad d_2 = d_1 - \sigma\sqrt{T}.$$

Therefore,

$$z_1 = d_1(x, K), \ z_4 = d_2(x, K)$$

$$\implies xN(z_1) - Ke^{-rT}N(z_4) = C(x, K),$$

$$z_2 = d_1(x, L), \ z_5 = d_2(x, L) \implies xN(z_2) - Ke^{-rT}N(z_5)$$

$$= \frac{K}{L}\left(\frac{xL}{K}N(z_2) - Le^{-rT}N(z_5)\right) = \frac{K}{L}C(xL/K, L),$$

$$z_3 = d_1(x, x^2/L), \ z_6 = d_2(x, x^2/L)$$

$$\implies x\left(\frac{L}{x}\right)^{\frac{2r}{\sigma^2}+1} N(z_3) - Ke^{-rT}\left(\frac{L}{x}\right)^{\frac{2r}{\sigma^2}-1} N(z_6)$$

$$= \left(\frac{L}{x}\right)^{\frac{2r}{\sigma^2}+1} \frac{K}{L}\left(\frac{xL}{K}N(z_3) - e^{-rT}\frac{x^2}{L}N(z_6)\right)$$

$$= \left(\frac{L}{x}\right)^{\frac{2r}{\sigma^2}+1} \frac{K}{L}C(xL/K, x^2/L)$$

$$z_7 = d_1(x, x^2K/L^2), \ z_8 = d_2(x, x^2K/L^2)$$

$$\implies x\left(\frac{L}{x}\right)^{\frac{2r}{\sigma^2}+1} N(z_7) - Ke^{-rT}\left(\frac{L}{x}\right)^{\frac{2r}{\sigma^2}-1} N(z_8)$$

$$= \left(\frac{L}{x}\right)^{\frac{2r}{\sigma^2}+1} \left(xN(z_7) - e^{-rT}\frac{Kx^2}{L^2}N(z_8)\right)$$

$$= \left(\frac{L}{x}\right)^{\frac{2r}{\sigma^2}+1} C(x, Kx^2/L^2).$$

Thus, we get the following expression:

$$DIC(x, K, L) = \begin{cases} C(x, K) - \dfrac{K}{L}C(xL/K, L) \\ \quad + \left(\dfrac{L}{x}\right)^{\frac{2r}{\sigma^2}+1}\dfrac{K}{L}C(xL/K, x^2/L), \ x > L > K, \\ \left(\dfrac{L}{x}\right)^{\frac{2r}{\sigma^2}+1} C(x, Kx^2/L^2), \quad x > L, \ K > L, \\ C(x, K), \hspace{4.5cm} x \leqslant L. \end{cases} \qquad [2.40]$$

This and similar formulas allow us to construct replicating strategies for barrier options using European options.

Similarly, we obtain the price of the DOC option (with z_i, $i = 1, \ldots, 8$, defined by equations [2.38a–d]:

$$DOC(x, K, L) = \begin{cases} x\left[\mathcal{N}(z_2) - \left(\dfrac{L}{x}\right)^{\frac{2r}{\sigma^2}+1}\mathcal{N}(z_3)\right] \\ \quad - Ke^{-rT}\left[\mathcal{N}(z_5) - \left(\dfrac{L}{x}\right)^{\frac{2r}{\sigma^2}-1}\mathcal{N}(z_6)\right] \\ \qquad\qquad\qquad\qquad \text{for } x > L > K, \\ x\left[\mathcal{N}(z_1) - \left(\dfrac{L}{x}\right)^{\frac{2r}{\sigma^2}+1}\mathcal{N}(z_7)\right] \\ \quad - Ke^{-rT}\left[\mathcal{N}(z_4) - \left(\dfrac{L}{x}\right)^{\frac{2r}{\sigma^2}-1}\mathcal{N}(z_8)\right] \\ \qquad\qquad\qquad\qquad \text{for } x > L,\ K \geqslant L, \\ 0 \qquad\qquad\qquad\qquad\ \text{for } x \leqslant L. \end{cases}$$

[2.41]

If we know the price of one of the DIC and DOC options, the other price can be calculated by applying relation [2.34], as, in fact, we did when plotting the graphs of option prices in Figures 2.8 and 2.9.

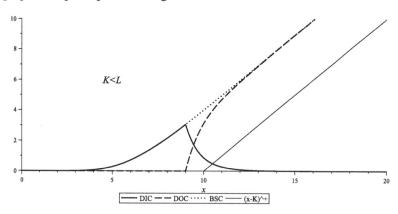

Figure 2.8. *Graphs of* $DIC(x, K, L)$, $DOC(x, K, L)$, *and* $C(x, K)$
when $K < L$ $(K = 10,\ L = 11)$. $r = 0.05; \sigma = 0.1; T = 10$.
$DIC(x, K, L) = C(x, K)$ *for* $x \leqslant L$

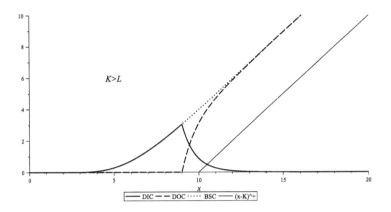

Figure 2.9. *Graphs of $DIC(x, K, L)$, $DOC(x, K, L)$, and $C(x, K)$*
when $K > L$ $(K = 10, L = 9)$. $r = 0.05; \sigma = 0.1; T = 10$.
$DIC(x, K, L) = C(x, K)$ for $x \leqslant L$

We can also similarly calculate the prices of the UOC and UIC options (with z_i, $i = 1, \ldots, 8$, defined by the same equations [2.38a–d], but with the barrier L replaced by the barrier H):

$$UOC(x, K, H) = \begin{cases} x\left[\mathcal{N}(z_1) - \mathcal{N}(z_2) + \left(\dfrac{H}{x}\right)^{\frac{2r}{\sigma^2}+1}(\mathcal{N}(z_3) - \mathcal{N}(z_7))\right] \\ \quad - Ke^{-rT}\left[\mathcal{N}(z_4) - \mathcal{N}(z_5)\right. \\ \qquad\qquad \left. + \left(\dfrac{H}{x}\right)^{\frac{2r}{\sigma^2}-1}(\mathcal{N}(z_6) - \mathcal{N}(z_8))\right] \\ \qquad\qquad\qquad\qquad \text{if } x \vee K < H, \\ 0 \qquad\qquad\qquad\qquad \text{if } x \vee K \geqslant H, \end{cases} \quad [2.42]$$

$$UIC(x, K, H) = \begin{cases} x\left[\mathcal{N}(z_2) - \left(\dfrac{H}{x}\right)^{\frac{2r}{\sigma^2}+1}(\mathcal{N}(z_3) - \mathcal{N}(z_7))\right] \\ - Ke^{-rT}\left[\mathcal{N}(z_5) - \left(\dfrac{H}{x}\right)^{\frac{2r}{\sigma^2}-1}(\mathcal{N}(z_6) - \mathcal{N}(z_8))\right] \\ \qquad\qquad\qquad\qquad \text{if } x \vee K < H, \\ C(x, K) \qquad\qquad\qquad \text{if } x \vee K \geqslant H. \end{cases} \quad [2.43]$$

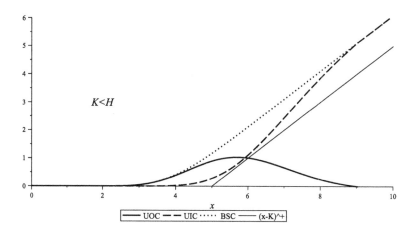

Figure 2.10. *Graphs of* $UIC(x, K, H)$, $UOC(x, K, H)$, *and* $C(x, K)$
when $K < H$ *(K = 5, H = 9)*; $r = 0.05$; $\sigma = 0.1$; $T = 5$

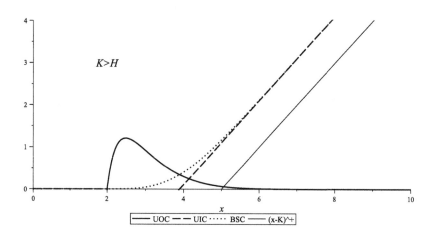

Figure 2.11. *Graphs of* $UIC(x, K, H)$, $UOC(x, K, H)$, *and* $C(x, K)$
when $K > H$ *(K = 5, H = 2)*; $r = 0.05$; $\sigma = 0.1$; $T = 5$

The prices of the barrier put options DIP, DOP, UIP and UOP can be calculated similarly. The calculations can be somewhat simplified by using the formulas obtained for barrier call options and the obvious relations between call and put options, for example,

$$DIC(x, K, L) - DIP(x, K, L) = \widetilde{\mathbf{E}}\big[e^{-rT}(S_T - K)\mathbf{1}_{\{\min_{t \leqslant T} S_t \leqslant L\}}\big].$$

The expectation on the right-hand side, where instead of the previous $(S_T - K)^{\pm}$ we have $S_T - K$, can be calculated relatively more simply.

2.7.4. *Lookback option pricing*

As an example, let us look for the price of the lookback call (LBC) option:

$$LBC(S_0) = \mathrm{e}^{-rT}\,\widetilde{\mathbf{E}}(S_T - \min_{t\in[0,T]} S_t)^+ = \mathrm{e}^{-rT}\,\widetilde{\mathbf{E}}(S_T - \min_{t\in[0,T]} S_t)$$

$$= \widetilde{\mathbf{E}}\widetilde{S}_T - \mathrm{e}^{-rT}\,\widetilde{\mathbf{E}}\min_{t\in[0,T]} S_0\exp\{\sigma\widetilde{B}_t + (r - \sigma^2/2)t\}$$

$$= S_0 - \mathrm{e}^{-rT}S_0\widetilde{\mathbf{E}}\exp\left\{\sigma\min_{t\in[0,T]}\widehat{B}_t\right\} \qquad [2.44]$$

with $\widehat{B}_t = \widetilde{B}_t + \mu t$, where $\mu = \sigma^{-1}(r - \sigma^2/2)$. To calculate the second term, as in section 2.7.3 (p. 69), we change the probability $\widetilde{\mathbf{P}}$ by the equivalent probability $\widehat{\mathbf{P}}$ such that $\widehat{B}$ becomes a Brownian motion with respect to it. Then (after taking off the hats $\widehat{}$ from $\widehat{\mathbf{P}}$, $\widehat{B}_t$, and $\widehat{B}_T$), we arrive at the equality

$$\widetilde{\mathbf{E}}\exp\left\{\sigma\min_{t\in[0,T]}\widehat{B}_t\right\} = \mathbf{E}\left(\exp\left\{\sigma\min_{t\in[0,T]}B_t\right\}\exp\left\{\mu B_T - \frac{\mu^2 T}{2}\right\}\right)$$

$$= \exp\{-\mu^2 T/2\}\mathbf{E}\exp\left\{\sigma m_T + \mu B_T\right\}$$

where $m_T := \min_{t\in[0,T]} B_t$. Recalling the joint density of (B_T, m_T) (see equation [1.2]),

$$p_T(x,y) = \frac{2(x - 2y)}{\sqrt{2\pi T^3}}\mathrm{e}^{-(x-2y)^2/2T}, \quad y \leqslant x \wedge 0,$$

we have

$$\mathbf{E}\exp\left\{\sigma m_T + \mu B_T\right\} = \int_{-\infty}^{0} \mathrm{d}y \int_{y}^{\infty} \exp\{\sigma y + \mu x\}p_T(x,y)\,\mathrm{d}x$$

$$= \frac{2}{\sqrt{2\pi T^3}}\int_{-\infty}^{0} \mathrm{d}y \int_{y}^{\infty}(x - 2y)\exp\{\sigma y + \mu x - (x - 2y)^2/2T\}\,\mathrm{d}x.$$

Calculation of this integral is very tedious (although possible). Therefore, let us seek an appropriate formula in the literature. In [BOR 02], p. 250, we find formula [1.1.3] for $\mathbf{E}\exp\{-\gamma\max_{s\leqslant t}(B_s + \mu s)\}$. Noting that $\min_{s\leqslant t}(B_s) =$

$- \max_{s \leqslant t}(-B_s - \mu s)$ and $-B$ again is a Brownian motion, the formula can be rewritten, in our notation, as follows:

$$\widetilde{\mathbf{E}} \exp\left\{\sigma \min_{t \in [0,T]} \widehat{B}_t\right\} = \mathbf{E} \exp\left\{\sigma \min_{t \in [0,T]} (B_t + \mu t)\right\}$$

$$= \frac{\sigma + \mu}{\sigma + 2\mu} e^{\sigma(\sigma + 2\mu)T/2} \operatorname{Erfc}\left(\frac{(\sigma + \mu)\sqrt{T}}{\sqrt{2}}\right) + \frac{\mu}{\sigma + \mu} \operatorname{Erfc}\left(\frac{-\mu\sqrt{t}}{\sqrt{2}}\right), \qquad [2.45]$$

where

$$\operatorname{Erfc}(x) := \frac{2}{\sqrt{\pi}} \int_x^\infty e^{-t^2} \, dt = 2(1 - \mathcal{N}(x\sqrt{2})), \quad x \in \mathbb{R}, \qquad [2.46]$$

is the so-called complementary error function. Making the substitutions [2.46],

$$\mu = \frac{r - \sigma^2/2}{\sigma}, \quad \sigma + \mu = \frac{r + \sigma^2/2}{\sigma}, \quad \text{and} \quad \sigma + 2\mu = \frac{2r}{\sigma}$$

into equation [2.45] and then into equation [2.44], we get the final formula for $LBC(S_0)$:

$$LBC(x) = x\left[\mathcal{N}(d_1) - e^{-rT}\mathcal{N}(d_2)\right.$$

$$\left. + \frac{\sigma^2}{2r}\left(e^{-rT}\mathcal{N}(d_2) - \mathcal{N}(-d_1)\right)\right] \qquad [2.47]$$

with

$$d_1 = \frac{(r + \sigma^2/2)\sqrt{T}}{\sigma}, \quad d_2 = \frac{(r - \sigma^2/2)\sqrt{T}}{\sigma} = d_1 - \sigma\sqrt{T}.$$

Formula [2.47] for the lookback call option price at the initial time moment $t = 0$ is relatively simple, since, at that moment, we have $S_0 = m_0$. We omit yet more tedious calculations for $0 < t \leqslant T$ and give only the final formula for $V_t = LBC(S_t, m_t)$:

$$LBC(x, m) = x\mathcal{N}(d_1) - me^{-r\tau}\mathcal{N}(d_2)$$

$$+ \frac{\sigma^2}{2r}\left[e^{-r\tau}x\left(\frac{m}{x}\right)^{2r/\sigma^2}\mathcal{N}(d_3) - \mathcal{N}(-d_1)\right] \qquad [2.48]$$

with $\tau = T - t$ and

$$d_1 = \frac{\ln(\frac{x}{m}) + (r + \sigma^2/2)\tau}{\sigma\sqrt{\tau}}, \quad d_2 = d_1 - \sigma\sqrt{\tau}, \quad d_3 = -d_1 + \frac{2r\sqrt{\tau}}{\sigma}.$$

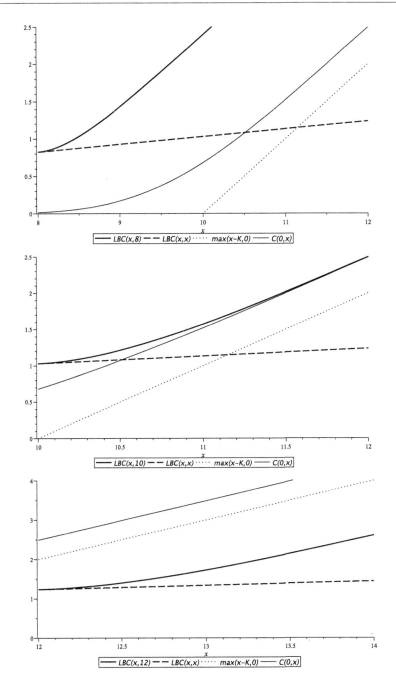

Figure 2.12. *Comparison of lookback and vanilla call option prices*
$LBC(x, m)$, $LBC(x, x)$ *and* $C(x, K)$ *in the cases* $m < K$, $m = K$ *and*
$m > K$: $m = 8, 10, 12$; $K = 10$; $r = 0.05$; $\sigma = 0.1$; $\tau = 1$

Note that if $x = m$, then $d_3 = d_2$, and formula [2.48] coincides with formula [2.47].

Let us try to compare the prices of lookback and vanilla options, $LBC(x, m)$ and $C(x.K)$. Note that the price $LBC(x, m)$ makes sense only for $x \geqslant m$. In Figure 2.12, we see typical plots of $LBC(x, m)$ (equation [2.48]), $LBC(x)$ (equation [2.47]) and $C(0, x)$ (equation [2.18]) in the cases $m < K$, $m = K$, and $m > K$.

Models of Interest Rates

3.1. Modeling principles

Until now, in all the models considered, the interest rates were constant. In reality, the interest rates are random and they depend on both initial time moment t and maturity T. Suppose that you loan some money (say, by buying a bond from the government) at moment t that will be repaid by one dollar at maturity T. This is the so-called zero-coupon bond, a security that provides no other cash flows between times t and T (i.e. it does not make periodic "coupon" payments, hence the term zero-coupon bond). In other words, at time moment t, you buy a bond for a certain price $P(t, T)$ the value of which at maturity T is 1. How can we calculate this price? What must be the properties of the function $P(t, T)$ in an arbitrage-free market? First, it obviously must be positive and satisfy the end condition $P(T, T) = 1$. If it is "too high," you will prefer to invest your money elsewhere. If it is "too low," the lender will suffer a loss because of his commitment being greater than necessary.

In the simplest case where the interest rate is constant and equals r in the time interval $[0, T]$, it is clear that the "fair" price is $P(t, T) = e^{-(T-t)r}$. This motivates the following definition in the general case. The average interest rate $R(t, T)$ in the interval $[t, T]$ is defined by the equality

$$P(t, T) = e^{-(T-t)R(t,T)},$$

i.e.

$$R(t, T) = -\frac{1}{(T - t)} \ln P(t, T).$$

The function $R(t, T)$ as a function of the variable $T \geqslant t$ is called the *yield curve* (yield to maturity T at time t, often denoted $Y(t, T)$). The function

$$f(t, T) = -\frac{\partial \ln P(t, T)}{\partial T}$$

is called the *forward spot rate* (at time t with maturity T). It is interpreted as the spot rate at time T as it is seen at time t (i.e. in a sense, the expected *future spot rate*). Let us comment on this in more detail. By integrating the last equality, we have

$$\int_t^T f(t, s)\, ds = -\int_t^T \frac{\partial \ln P(t, s)}{\partial s}\, ds = -\ln P(t, s)\Big|_{s=t}^T$$
$$= -\ln P(t, T),$$

that is,

$$P(t, T) = \exp\left\{ -\int_t^T f(t, s)\, ds \right\}.$$ [3.1]

Now suppose that after buying a zero-coupon bond for $P(t, T)$ dollars, we reinvest it with the agreement that at each future moment $s \in [t, T]$, we will be paid dividends with interest rate $f(t, s)$. Then, at the moment T, we will be paid

$$P(t, T) \exp\left\{ \int_t^T f(t, s)\, ds \right\} = 1 \text{ dollar},$$

i.e. the same sum as we would get without reinvestment of bond. This explains the term *forward spot rate* (or the expected future spot rate).

If the future is known, i.e. all average interest rates $R(t, T)$ are known, then, in the arbitrage-free market, the price function P must satisfy the condition

$$P(t, s) = P(t, u)P(u, s), \quad t \leqslant u \leqslant s.$$ [3.2]

PROPOSITION 3.1.– If a continuously differentiable function $P(t,s)$, $t_0 \leqslant t \leqslant s \leqslant T$, satisfies the multiplicativity condition [3.2] and $P(t,t) \equiv 1$, then there exists a function r such that

$$P(t,s) = \exp\left\{ -\int_t^s r(u)\,du \right\}, \quad t_0 \leqslant t \leqslant s \leqslant T,$$

i.e. the average interest rate $R(t,s) = \frac{1}{s-t}\int_t^s r(u)\,du$. The function r is called the instantaneous interest rate (or the spot interest rate).

PROOF.– Denote $F = P^{-1}$ and $r(u) = F_u'(t,u)|_{t=u}$. Then,

$$r(u) = \lim_{h\downarrow 0} \frac{F(u,u+h) - F(u,u)}{h} = \lim_{h\downarrow 0} \frac{F(t,u+h)/F(t,u) - 1}{h}$$

$$= \frac{1}{F(t,u)} \lim_{h\downarrow 0} \frac{F(t,u+h) - F(t,u)}{h} = \frac{F_u'(t,u)}{F(t,u)} = (\ln F(t,u))_u'$$

$$\Longrightarrow \ln F(t,s) - \ln F(t,t) = \int_t^s r(u)\,du$$

$$\Longrightarrow P(t,s) = F^{-1}(t,s) = \exp\left\{ -\int_t^s r(u)\,du \right\}.$$

From this, we get a formula for the average interest rate:

$$R(t,T) = \frac{1}{T-t}\int_t^T r(s)\,ds, \quad t_0 \leqslant t \leqslant T, \tag{3.3}$$

In the real world, at time t, average interest rates $R(u,T)$ at future time moments $u \in (t,T)$ are not known, and in equation [3.3], r is an (adapted) random process. It is also assumed that, in the market, together with the "risky" bond whose price at moment t is $P(t,T)$, there is the "riskless" investment whose price at moment t is

$$S_t^0 = \exp\left\{ \int_0^t r_s\,ds \right\}.$$

Based on the no-arbitrage principle, there exists a risk-neutral probability $\widetilde{\mathbf{P}}$ such that the discounted prices of the bond,

$$P(t,T)/S_t^0, \quad t \in [0,T],$$

are $\widetilde{\mathbf{P}}$-martingales for all T. From this, we get

$$P(t,T)/S_t^0 = \widetilde{\mathbf{E}}[P(T,T)/S_T^0|\mathcal{F}_t] = \widetilde{\mathbf{E}}[1/S_T^0|\mathcal{F}_t], \quad t \in [0,T].$$

Therefore,

$$P(t,T) = S_t^0 \widetilde{\mathbf{E}}[1/S_T^0|\mathcal{F}_t] = \exp\left\{\int_0^t r_s \, ds\right\} \widetilde{\mathbf{E}}\left[\exp\left\{-\int_0^T r_s \, ds\right\}\bigg|\mathcal{F}_t\right]$$

$$= \widetilde{\mathbf{E}}\left[\exp\left\{-\int_t^T r_s \, ds\right\}\bigg|\mathcal{F}_t\right]. \tag{3.4}$$

If the forward spot rate $f(t,T)$ is known, we can easily recover the spot interest rate r_t. Differentiating equations [3.4] and [3.2] with respect to T, we have

$$\frac{\partial P(t,T)}{\partial T} = \widetilde{\mathbf{E}}\left[-r_T \exp\left\{-\int_t^T r_s \, ds\right\}\bigg|\mathcal{F}_t\right] \Longrightarrow$$

$$\frac{\partial P(t,T)}{\partial T}\bigg|_{T=t} = \widetilde{\mathbf{E}}\left[-r_t|\mathcal{F}_t\right] = -r_t,$$

and

$$\frac{\partial P(t,T)}{\partial T} = -f(t,T) \exp\left\{-\int_t^T f(t,s) \, ds\right\} \Longrightarrow$$

$$\frac{\partial P(t,T)}{\partial T}\bigg|_{T=t} = -f(t,t),$$

i.e.

$$r_t = f(t,t). \tag{3.5}$$

There are two ways of modeling interest rates. The first is based on modeling spot rate process r. The second is based on a direct modeling of prices $P(t,T)$ taking into account the no-arbitrage condition.

3.1.1. *Classic models of interest rates*

In such models, it is assumed that the spot rate process r satisfies the stochastic differential equation

$$dr_t = \alpha(t, r_t) \, dt + \sigma(t, r_t) \, dB_t.$$

To find the value of zero-coupon bond, we have to know a stochastic differential equation with respect to the risk-neutral probability $\widetilde{\mathbf{P}}$. Suppose that λ is a random process for which $\widetilde{B}_t = B_t + \int_0^t \lambda_s \, ds$ is a Brownian motion with respect to $\widetilde{\mathbf{P}}$ (see the Girsanov theorem). In classic models, the process λ usually takes the form $\lambda_t = \lambda(t, r_t)$. Then, the spot interest rate r satisfies the following SDE with respect to $\widetilde{\mathbf{P}}$:

$$\mathrm{d}r_t = \big(\alpha(t, r_t) - \lambda(t, r_t)\sigma(t, r_t)\big)\, \mathrm{d}t + \sigma(t, r_t)\mathrm{d}\widetilde{B}_t, \qquad [3.6]$$

i.e. the equation for r is of the same form with respect to both probabilities $\mathbf{P}$ and $\widetilde{\mathbf{P}}$. The process λ is called the *risk premium*.

As, under wide conditions, a solution of SDE [3.6] is a Markov process, formula [3.4] for the price $P(t, T)$ becomes

$$P(t, T) = \widetilde{\mathbf{E}}\left[\exp\left\{-\int_t^T r_s \, ds\right\} \bigg| r_t\right] = p(t, r_t, T), \qquad [3.7]$$

where

$$p(t, r, T) = \widetilde{\mathbf{E}}\left[\exp\left\{-\int_t^T r_s \, ds\right\} \bigg| r_t = r\right], \; 0 \leqslant t \leqslant T, \; r \in \mathbb{R}_+. [3.8]$$

3.1.2. *Partial differential equation for interest rates*

Let, as in the previous subsection, the value of a zero-coupon bond be of the form $P(t, T) = p(t, r_t, T)$ (equations [3.7]–[3.8]). Suppose that p is a sufficiently smooth function, namely, $p \in C^2([0, T] \times \mathbb{R}_+)$ for a fixed T. By Itô's formula (for fixed T),

$$\begin{aligned}
\mathrm{d}P(t, T) &= \frac{\partial p}{\partial t}(t, r_t, T)\, \mathrm{d}t + \frac{\partial p}{\partial r}(t, r_t, T)\mathrm{d}r_t + \frac{1}{2}\frac{\partial^2 p}{\partial r^2}(t, r_t, T)\mathrm{d}\langle r \rangle_t \\
&= \left(\frac{\partial p}{\partial t} + \alpha\frac{\partial p}{\partial r} + \frac{\sigma^2}{2}\frac{\partial^2 p}{\partial r^2}\right)(t, r_t, T)\, \mathrm{d}t + \sigma\frac{\partial p}{\partial r}(t, r_t, T)\, \mathrm{d}B_t \\
&= P(t, T)\left(\mu(t, r_t, T)\, \mathrm{d}t + \gamma(t, r_t, T)\, \mathrm{d}B_t\right) \qquad [3.9]
\end{aligned}$$

with

$$\mu(t,r,T) = \frac{1}{p(t,r,T)} \left(\frac{\partial p}{\partial t} + \alpha \frac{\partial p}{\partial r} + \frac{\sigma^2}{2} \frac{\partial^2 p}{\partial r^2} \right) (t,r,T),$$

$$\gamma(t,r,T) = \frac{1}{p(t,r,T)} \sigma(t,r) \frac{\partial p}{\partial r}(t,r,T).$$

Equation [3.9] is obtained with respect to $\mathbf{P}$ (where B is a Brownian motion with respect to $\mathbf{P}$). Replacing the latter by $\widetilde{\mathbf{P}}$, the equation can be written as

$$dP(t,T) = P(t,T) \Big[(\mu(t,r_t,T) - \lambda(t,r_t)\gamma(t,r_t,T)) \, dt$$

$$+ \gamma(t,r_t,T) \, d\widetilde{B}_t \Big], \qquad\qquad [3.10]$$

where $\widetilde{B}$ is a Brownian motion with respect to $\widetilde{\mathbf{P}}$. The discounted value $P/S = P(t,T)/S_t^0$ is a $\widetilde{\mathbf{P}}$-martingale. On the other hand,

$$d(P/S) = S^{-1}dP + P\,dS^{-1}$$

$$= S^{-1}P \Big[(\mu_t - \lambda_t\gamma_t) \, dt + \gamma_t \, d\widetilde{B}_t \Big] - PS^{-1}r_t \, dt$$

$$= S^{-1}P \left(\mu_t - \lambda_t\gamma_t - r_t \right) \, dt + S^{-1}P\gamma_t \, d\widetilde{B}_t.$$

As the last term is a differential of a $\widetilde{\mathbf{P}}$-martingale, the drift coefficient must equal zero. Thus,

$$\mu_t - \gamma_t\lambda_t = \mu(t,r_t,T) - \lambda(t,r_t)\gamma(t,r_t,T) = r_t. \qquad\qquad [3.11]$$

Substituting the expressions of μ and γ into the last equality, we obtain the following equation for the function p:

$$\frac{1}{p(t,r,T)} \left(\frac{\partial p}{\partial t} + \alpha \frac{\partial p}{\partial r} + \frac{\sigma^2}{2} \frac{\partial^2 p}{\partial r^2} \right) (t,r,T)$$

$$- \frac{1}{p(t,r,T)} \lambda(t,r)\sigma(t,r) \frac{\partial p}{\partial r}(t,r,T) = r.$$

Multiplying by $p = p(t,r,T)$, we finally get the partial differential equation

$$\boxed{\frac{\partial p}{\partial t} + (\alpha - \lambda\sigma) \frac{\partial p}{\partial r} + \frac{\sigma^2}{2} \frac{\partial^2 p}{\partial r^2} = r\,p} \qquad\qquad [3.12]$$

with end condition[1]

$$p(T, r, T) = 1. \tag{3.13}$$

REMARK 3.1.– 1) It is interesting to compare the partial differential equation [3.12] with the Black–Scholes equation [2.13]. The main difference is the appearance of the risk premium process λ. This parameter is not defined in the model but must be known a priori. Usually, it is estimated statistically on the basis of market prices.

2) As in the Black–Scholes model, for calculating the values of zero-coupon bonds in the model [3.6], two methods can be followed. The first is based on the probabilistic expression [3.4] (or [3.7] for a classic model) of the value in terms of risk-neutral probability, and the second is based on the partial differential equation [3.12]–[3.13].

3) From equations [3.10] and [3.11], we obtain the equation

$$\mathrm{d}P(t, T) = P(t, T)\big(r_t \,\mathrm{d}t + \gamma(t, r_t, T)\,\mathrm{d}\widetilde{B}_t\big). \tag{3.14}$$

We see that in the classic interest rate models, the mean rate of return with respect to $\widetilde{\mathbf{P}}$ is necessarily equal to the spot interest rate r_t.

EXERCISE 3.1.– Suppose that the spot rate is a Brownian motion with a drift, i.e.

$$\mathrm{d}r_t = \mu \,\mathrm{d}t + \sigma \,\mathrm{d}B_t.$$

Find the zero-coupon value

$$p(t, r, T) = \widetilde{\mathbf{E}}\left[\exp\left\{-\int_t^T r_s \,\mathrm{d}s\right\}\bigg|\, r_t = r\right], \quad 0 \leqslant t \leqslant T,\ r \in \mathbb{R},$$

and check that it satisfies the corresponding PDE [3.12].

3.2. The Vašíček model

In this model, the spot rates satisfy the Ornstein–Uhlenbeck equation

$$\mathrm{d}r_t = a(b - r_t)\,\mathrm{d}t + \sigma \,\mathrm{d}B_t,$$

---

1 Since $P(T, T) = 1$.

where a, b and σ are positive constants. The "force" proportional to the coefficient a "pushes" r_t toward the value b. It is also assumed that the risk premium λ is constant, i.e. the Brownian motion with drift $\widetilde{B}_t = B_t + \lambda t$ is a Brownian motion with respect to risk-neutral probability $\widetilde{\mathbf{P}}$. Therefore, r satisfies the equation

$$\mathrm{d}r_t = a(\tilde{b} - r_t)\,\mathrm{d}t + \sigma\mathrm{d}\widetilde{B}_t, \qquad\qquad [3.15]$$

where $\tilde{b} = b - \sigma\lambda/a$.

PROPOSITION 3.2.– The solution of equation [3.15] is

$$r_t = \tilde{b} + (r_0 - \tilde{b})\mathrm{e}^{-at} + \sigma \int_0^t \mathrm{e}^{-a(t-s)}\mathrm{d}\widetilde{B}_s. \qquad\qquad [3.16]$$

PROOF.–

Applying the integration-by-parts formula (theorem 1.18) to $r_t e^{at}$, we have

$$r_t e^{at} = r_0 + \int_0^t r_u \mathrm{d}e^{as} + \int_0^t e^{au}\mathrm{d}r_s$$

$$= r_0 + \int_0^t r_s a e^{as}\,\mathrm{d}u + \int_0^t e^{as}\left(a(\tilde{b} - r_s)\,\mathrm{d}s + \sigma\mathrm{d}\widetilde{B}_s\right)$$

$$= r_0 + a\tilde{b}\int_0^t a e^{as}\,\mathrm{d}s + \sigma\int_0^t e^{as}\mathrm{d}\widetilde{B}_s$$

$$= r_0 + \tilde{b}(e^{at} - 1) + \sigma\int_0^t e^{as}\mathrm{d}\widetilde{B}_s$$

and thus

$$r_t = r_0 e^{-at} + \tilde{b}(1 - e^{-at}) + \sigma e^{-at}\int_0^t e^{as}\mathrm{d}\widetilde{B}_s$$

$$= \tilde{b} + (r_0 - \tilde{b})\mathrm{e}^{-at} + \sigma \int_0^t \mathrm{e}^{-a(t-s)}\mathrm{d}\widetilde{B}_s. \qquad\qquad \square$$

Thus, we see that, with respect to $\widetilde{\mathbf{P}}$, r_t is a Gaussian process with mean

$$\widetilde{\mathbf{E}}r_t = \tilde{b} + (r_0 - \tilde{b})\mathrm{e}^{-at} \qquad\qquad [3.17]$$

and variance

$$\widetilde{\mathrm{Var}}\, r_t = \widetilde{\mathrm{E}} \left(\sigma \int_0^t \mathrm{e}^{-a(t-s)} \mathrm{d}\widetilde{B}_s \right)^2 = \sigma^2 \int_0^t \mathrm{e}^{-2a(t-s)}\, \mathrm{d}s$$

$$= \frac{\sigma^2}{2a}(1 - \mathrm{e}^{-2at}). \hspace{3cm} [3.18]$$

Note that r_t may take positive values, which is undesirable in a realistic model. Also note that, as $t \to \infty$, r_t converges in distribution to a normal random variable with mean $\tilde{b}$ (hence the name *mean-reverting* process) and variance $\sigma^2/2a$. To calculate the value of bond by formula [3.4], we need to know the distribution of the integral $\int_t^T r_s\, \mathrm{d}s$ given $\mathcal{F}_t$ or, equivalently, given r_t (since r is a Markov process). We find it by integrating equation [3.15] from t to T:

$$r_T - r_t = a\tilde{b}(T - t) - a \int_t^T r_s\, \mathrm{d}s + \sigma(\widetilde{B}_T - \widetilde{B}_t) \Longrightarrow$$

$$a \int_t^T r_s\, \mathrm{d}s = a\tilde{b}(T - t) - (r_T - r_t) + \sigma(\widetilde{B}_T - \widetilde{B}_t). \hspace{1cm} [3.19]$$

From expression [3.16] of r_t (by replacing the time interval $[0, t]$ by $[t, T]$), we have

$$r_T = \tilde{b} + (r_t - \tilde{b})\mathrm{e}^{-a(T-t)} + \sigma \int_t^T \mathrm{e}^{-a(T-s)}\, \mathrm{d}\widetilde{B}_s.$$

Substituting this into equation [3.19], we get

$$\int_t^T r_s\, \mathrm{d}s = \frac{1}{a} \left[a\tilde{b}(T - t) - (r_T - r_t) + \sigma \int_t^T \mathrm{d}\widetilde{B}_s \right]$$

$$= \tilde{b}(T - t) - \frac{1}{a} \left[\tilde{b} - r_t - (r_t - \tilde{b})\mathrm{e}^{-a(T-t)} \right.$$

$$\left. + \sigma \int_t^T (1 - \mathrm{e}^{-a(T-s)})\, \mathrm{d}\widetilde{B}_s \right]$$

$$= \tilde{b}(T - t) + (r_t - \tilde{b})\frac{1 - \mathrm{e}^{-a(T-t)}}{a}$$

$$+ \sigma \int_t^T \frac{1 - \mathrm{e}^{-a(T-s)}}{a}\, \mathrm{d}\widetilde{B}_s.$$

From this, we see that, given r_t, the integral $\int_t^T r_s \, ds$ is a normal random variable with the following conditional mean and variance:

$$E(t,T) = \tilde{\mathbf{E}}\left[\int_t^T r_s \, ds \middle| r_t\right] = \tilde{b}(T-t) + (r_t - \tilde{b})\frac{1 - e^{-a(T-t)}}{a}$$

[3.20]

and

$$V(t,T) = \widetilde{\operatorname{Var}}\left[\int_t^T r_s \, ds \middle| r_t\right] = \tilde{\mathbf{E}}\left[\left(\sigma \int_t^T \frac{1 - e^{-a(T-s)}}{a} \, d\tilde{B}_s\right)^2 \middle| r_t\right]$$

$$= \sigma^2 \int_t^T \left(\frac{1 - e^{-a(T-s)}}{a}\right)^2 \, ds$$

$$= \frac{\sigma^2}{a^2} \int_t^T \left(1 - 2e^{-a(T-s)} + e^{-2a(T-s)}\right) \, ds$$

$$= \frac{\sigma^2}{a^2}\left(s - \frac{2}{a}e^{-a(T-s)} + \frac{1}{2a}e^{-2a(T-s)}\right)\bigg|_{s=t}^T$$

$$= \frac{\sigma^2}{a^2}\left(T - t - \frac{2}{a}\left(1 - e^{-a(T-t)}\right) + \frac{1}{2a}\left(1 - e^{-2a(T-t)}\right)\right)$$

$$= \frac{\sigma^2}{a^2}(T-t) - \frac{2\sigma^2}{a^3}\left(1 - e^{-a(T-t)}\right) + \frac{\sigma^2}{2a^3}\left(1 - e^{-2a(T-t)}\right)$$

$$= \frac{\sigma^2}{a^2}(T-t) - \frac{\sigma^2}{2a^3}\left(1 - e^{-a(T-t)}\right)^2 - \frac{\sigma^2}{a^3}\left(1 - e^{-a(T-t)}\right).$$

[3.21]

THEOREM 3.1.– In the Vašíček model, the value of zero-coupon bond is

$$P(t,T) = \exp\left\{-R_\infty(T-t) + (R_\infty - r_t)\frac{1 - e^{-a(T-t)}}{a}\right.$$

$$\left. - \frac{\sigma^2}{4a^3}\left(1 - e^{-a(T-t)}\right)^2\right\},$$

[3.22]

where $R_\infty = \tilde{b} - \frac{\sigma^2}{2a^2} = b - \frac{\lambda\sigma}{a} - \frac{\sigma^2}{2a^2}$. The yield curve at moment t is

$$R(t,T) = R_\infty - (R_\infty - r_t)\frac{1 - e^{-a(T-t)}}{a(T-t)} + \frac{\sigma^2}{4a^3(T-t)}\left(1 - e^{-a(T-t)}\right)^2.$$

PROOF.– As $\int_t^T r_s \, ds$ is a normal random variable with conditional mean $E(t,T)$ and variance $V(t,T)$ with respect to σ-algebra $\mathcal{F}_t$, we have

$$
P(t,T) = \tilde{\mathbf{E}} \left[\exp \left\{ - \int_t^T r_s \, ds \right\} \middle| \mathcal{F}_t \right]
$$

$$
= \exp \left\{ -E(t,T) + \frac{1}{2} V(t,T) \right\},
$$

where we have used that the Laplace transform of a normal random variable $\xi \sim N(a, \sigma^2)$ is $\mathbf{E}e^{\lambda \xi} = e^{\lambda a + \lambda^2 \sigma^2 / 2}$. Taking $\lambda = -1$ and substituting expressions [3.20]–[3.21], we have

$$
p(t,T) = \exp \left\{ - \tilde{b}(T-t) - (r_t - \tilde{b}) \frac{1 - e^{-a(T-t)}}{a} \right.
$$

$$
\left. + \frac{1}{2} \left(\frac{\sigma^2}{a^2}(T-t) - \frac{\sigma^2}{2a^3} \left(1 - e^{-a(T-t)}\right)^2 - \frac{\sigma^2}{a^3} \left(1 - e^{-a(T-t)}\right) \right) \right\}
$$

$$
= \exp \left\{ - \left(\tilde{b} - \frac{\sigma^2}{2a^2}\right)(T-t) + \left(\tilde{b} - \frac{\sigma^2}{2a^2} - r_t\right) \frac{1 - e^{-a(T-t)}}{a} \right.
$$

$$
\left. - \frac{\sigma^2}{4a^3} \left(1 - e^{-a(T-t)}\right)^2 \right\},
$$

which is, in fact, formula [3.22]. From this we immediately obtain the formula for the yield curve $R(t,T) = -\frac{1}{T-t} \ln P(t,T)$.

Second way. We derived the formula for bond price on the basis of formula [3.4]. We shall try an alternative way based on the partial differential equation [3.12], which in this case becomes

$$
\frac{\partial p}{\partial t} + \left(a(b-r) - \lambda \sigma\right) \frac{\partial p}{\partial r} + \frac{\sigma^2}{2} \frac{\partial^2 p}{\partial r^2} = r \, p. \tag{3.23}
$$

Motivated by the multiplicativity property [3.2] of $P(t,T) = p(t, r_t, T)$, let us look for its solution in the form of a multiplicative function

$$
p(t, r, T) = \exp\{c(\theta)r + d(\theta)\} \tag{3.24}
$$

with coefficients $c = c(\theta)$ and $c = d(\theta)$ depending on $\theta := T - t$. Then, equation [3.23] becomes

$$-(c'(\theta)r + d'(\theta))p + (a(b - r) - \lambda\sigma)c(\theta)p + \frac{\sigma^2}{2}c^2(\theta)p = rp$$

with end condition $p(T, r, T) = 1$. Dividing by p and equating the terms at r and $r^0 = 1$, we get the system

$$\begin{cases} c'(\theta) + ac(\theta) + 1 = 0, \\ -d'(\theta) + (ab - \lambda\sigma)c(\theta) + \frac{\sigma^2}{2}c^2(\theta) = 0, \end{cases}$$

with initial conditions $c(0) = d(0) = 0$. Solving the latter, we get

$$c(\theta) = \frac{1}{a}\left(e^{-a\theta} - 1\right)$$

and

$$d(\theta) = d(0) + \int_0^\theta d'(\tau)d\tau = \int_0^\theta \left((ab - \lambda\sigma)c(\tau) + \frac{\sigma^2}{2}c^2(\tau)\right)d\tau$$

$$= \int_0^\theta \left[\left(b - \frac{\lambda\sigma}{a}\right)\left(e^{-a\tau} - 1\right) + \frac{\sigma^2}{2a^2}\left(e^{-a\tau} - 1\right)^2\right]d\tau$$

$$= \int_0^\theta \left[-\left(b - \frac{\lambda\sigma}{a} - \frac{\sigma^2}{2a^2}\right) + \left(b - \frac{\lambda\sigma}{a} - \frac{\sigma^2}{a^2}\right)e^{-a\tau} + \frac{\sigma^2}{2a^2}e^{-2a\tau}\right]d\tau$$

$$= -\left(b - \frac{\lambda\sigma}{a} - \frac{\sigma^2}{2a^2}\right)\theta + \left(b - \frac{\lambda\sigma}{a} - \frac{\sigma^2}{a^2}\right)\frac{1 - e^{-a\theta}}{a} + \frac{\sigma^2}{4a^3}(1 - e^{-2a\tau}).$$

Denoting $R_\infty = b - \frac{\lambda\sigma}{a} - \frac{\sigma^2}{2a^2}$, we arrive at

$$d(\theta) = -R_\infty\theta - \left(R_\infty - \frac{\sigma^2}{2a^2}\right)\frac{1 - e^{-a\theta}}{a} - \frac{\sigma^2}{4a^3}(1 - e^{-2a\theta})$$

$$= -R_\infty\theta + R_\infty\frac{1 - e^{-a\theta}}{a} - \frac{\sigma^2}{2a^3}(1 - e^{-a\theta}) + \frac{\sigma^2}{4a^3}(1 - e^{-2a\theta})$$

$$= -R_\infty\theta + R_\infty\frac{1 - e^{-a\theta}}{a} - \frac{\sigma^2}{4a^3}(1 - 2e^{-a\theta} + e^{-2a\theta})$$

$$= -R_\infty\theta + R_\infty\frac{1 - e^{-a\theta}}{a} - \frac{\sigma^2}{4a^3}(1 - e^{-a\theta})^2.$$

Substituting these expressions of c and d and $\theta = T - t$ into equation [3.24], we get again formula [3.22].

REMARK 3.2.– We obtained an explicit expression of the yield curve $T \mapsto R(t, T)$ in the Vašíček model for any t. Note that, for all t,

$$R_\infty = \lim_{T \to \infty} R(t, T).$$

Therefore, R_∞ can be interpreted as a long-term interest rate. It does not depend on the spot rate r_t. Analyzing the graph of the function $T \mapsto R(t, T)$, we can see that:

1) if $r_t < R_\infty - \sigma^2/4a^2$, the function strictly increases;

2) if $R_\infty - \sigma^2/4a^2 \leqslant r_t \leqslant R_\infty + \sigma^2/2a^2$, then initially the function increases and later decreases;

3) if $R_\infty + \sigma^2/2a^2 < r_t$, the function strictly decreases.

According to practitioners, similar yield curves are observable in real financial markets, though not always. Real yield curves for $r_t > R_\infty$ sometimes have "pits" that are impossible in the Vašíček model.

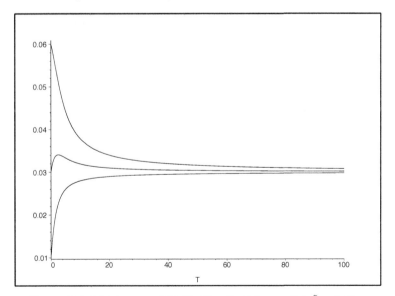

Figure 3.1. *Yield curves* $R(0, T)$, $T \in [0; 100]$: $\sigma = 0.1$; $\tilde{b} = 0.05$; $a = 0.5$; $R_\infty = 0.03$; $r_0 = 0.01$; 0.03; 0.06

EXERCISE 3.2.– The half-life of the Vašíček model $\mathrm{d}r_t = a(b-r_t)\,\mathrm{d}t + \sigma\,\mathrm{d}B_t$ is defined as the time it takes on average before the deviation from the long-term mean b is halved in size, i.e. the time $t_{1/2}$ such that

$$\mathbf{E}(r_{t_{1/2}} - b) = \frac{1}{2}(r_0 - b).$$

It shows the reversion rate of the Vašíček model to its long-term mean. Calculate the half-life $t_{1/2}$.

EXERCISE 3.3.– Show properties (1)–(3) of the yield curve stated at the end of this section.

EXERCISE 3.4.– (Exponentiated Vašíček model) Suppose that the logarithm of the sport rate follows an Ornstein–Uhlenbeck process, i.e. it satisfies the equation

$$\mathrm{d}r_t = ar_t(b - \ln r_t)\,\mathrm{d}t + \sigma r_t\,\mathrm{d}B_t,$$

and that the market price of risk is of the form $\lambda_t = \lambda \ln r_t + c$.

Write and solve the equation for r_t with Brownian motion $\tilde{B}$ with respect to risk-neutral probability $\tilde{\mathbf{P}}$. Find the mean $\tilde{\mathbf{E}}r_t$, second moment $\tilde{\mathbf{E}}r_t^2$, variance $\tilde{\mathbf{D}}r_t$, long-term mean $\tilde{\mathbf{E}}r_\infty := \lim_{t\to\infty}\tilde{\mathbf{E}}r_t$ and long-term variance $\tilde{\mathbf{D}}r_\infty := \lim_{t\to\infty}\tilde{\mathbf{D}}r_t$.

Further, consider a generalization of the Vašíček model, the Hull-White model

$$\mathrm{d}r_t = a(t)(b(t) - r_t)\,\mathrm{d}t + \sigma(t)\,\mathrm{d}B_t,$$

where a, b and σ are non-random positive functions. Assume that the risk premium $\lambda = \lambda(t)$ also is a non-random function, so that the Brownian motion with drift $\tilde{B}_t = B_t + \int_0^t \lambda(s)\,\mathrm{d}s$ is a Brownian motion with respect to risk-neutral probability $\tilde{\mathbf{P}}$.

EXERCISE 3.5.– Show that r satisfies the equation

$$\mathrm{d}r_t = a(t)(\tilde{b}(t) - r_t)\,\mathrm{d}t + \sigma(t)\,\mathrm{d}\tilde{B}_t, \qquad [3.25]$$

where $\tilde{b}(t) = b(t) - \sigma(t)\lambda(t)/a(t)$, and $\tilde{B}$ is a Brownian motion with respect to $\tilde{\mathbf{P}}$.

EXERCISE 3.6.– Denoting $F(t) = \int_0^t a(s)\,ds$, show that the solution of equation [3.25] is

$$r_t = e^{-F(t)}\left(r_0 + \int_0^t e^{F(s)}\left(a(s)\tilde{b}(s)\,ds + \sigma(s)d\tilde{B}_s\right)\right). \qquad [3.26]$$

EXERCISE 3.7.– Show that $P(t,T) = p(t, r_t, T)$ must satisfy

$$\frac{\partial p}{\partial t} + \left(a(t)(b(t) - r) - \lambda(t)\sigma(t)\right)\frac{\partial p}{\partial r} + \frac{1}{2}\sigma^2(t)\frac{\partial^2 p}{\partial r^2} = r\,p.$$

EXERCISE 3.8.– Show that $\int_0^t r_s\,ds$ is a normal variable with mean

$$E(T) = \int_0^T e^{-F(s)}\left(r_0 + \int_0^s e^{F(u)}a(u)b(u)\,du\right)dt$$

and variance

$$D(T) = \int_0^T e^{2F(s)}\sigma^2(s)\left(\int_s^T e^{-F(u)}\,du\right)^2 ds.$$

EXERCISE 3.9.– Show that the price $P(0,T)$ of zero-coupon bond is

$$P(0,T) = \mathbf{E}\exp\left\{-\int_0^T r_s\,ds\right\} = \exp\left\{-E(T) + \tfrac{1}{2}D(T)\right\}$$
$$= \exp\{-r_0C(T) - A(T)\},$$

where

$$C(T) = \int_0^T e^{-F(s)}\,ds,$$

$$A(T) = \int_0^T \int_0^s e^{-F(s)+F(u)}a(u)b(u)\,du\,dt$$
$$- \tfrac{1}{2}\int_0^T e^{2F(s)}\sigma^2(s)\left(\int_s^T e^{-F(u)}\,du\right)^2 ds.$$

EXERCISE 3.10.– Show that the price $P(t,T)$ of zero-coupon bond is

$$P(t,T) = \exp\{-r_0C(t,T) - A(t,T)\},$$

where

$$C(t,T) = e^{F(t)} \int_t^T e^{-F(s)} \, ds,$$

$$A(t,T) = \int_t^T \left[e^{F(s)} a(s) b(s) \left(\int_s^T e^{-F(u)} \, du \right) \right.$$
$$\left. - \tfrac{1}{2} e^{2F(s)} \sigma^2(s) \left(\int_s^T e^{-F(u)} \, du \right)^2 \right] ds.$$

3.3. The Cox–Ingersoll–Ross model

Trying to avoid negative values in the Vašíček model, Cox, Ingersoll and Ross [COX 85] proposed a "square root" model, later called the CIR model, where the spot rates are modeled by the equation

$$dr_t = a(b - r_t) \, dt + \sigma \sqrt{r_t} \, dB_t. \tag{3.27}$$

with $a, \sigma > 0$, $b \in \mathbb{R}$.

To better clarify the properties of the solution of this equation, we rewrite it in a slightly more convenient form:

$$dX_t = (a - bX_t) \, dt + \sigma \sqrt{X_t} \, dB_t. \tag{3.28}$$

THEOREM 3.2.– For every $x \geqslant 0$, equation [3.28] has a unique solution $X_t = X_t^x$, $t \geqslant 0$, with initial conditions $X_0 = x$. Moreover, $X_t \geqslant 0$ for all $t \geqslant 0$.

Denote $\tau_0^x = \inf\{t \geqslant 0 \colon X_t^x = 0\}$ ($\inf \emptyset = \infty$).

1) If $a \geqslant \sigma^2/2$, then $\mathbf{P}\{\tau_0^x = \infty\} = 1$ for all $x > 0$.

2) If $0 \leqslant a < \sigma^2/2$ and $b \geqslant 0$, then $\mathbf{P}\{\tau_0^x < \infty\} = 1$ for all $x > 0$.

3) If $0 \leqslant a < \sigma^2/2$ and $b < 0$, then $\mathbf{P}\{\tau_0^x < \infty\} \in (0,1)$ for all $x > 0$.

PROOF.– We omit the proof of the existence and uniqueness. We only prove properties (1)–(2) (following the scheme of exercise 34 in [LAM 96], section 6). Let

$$s(x) := \int_1^x y^{-2a/\sigma^2} e^{\frac{2b}{\sigma^2} y} \, dy, \quad x > 0.$$

We can directly check that the function s satisfies the equation

$$As(x) := (a - bx)s'(x) + \frac{\sigma^2}{2}xs''(x) = 0, \quad x > 0.$$

Denote $\tau_M^x := \inf\{t \geqslant 0 \colon X_t^x = M\}$ for all $x, M > 0$ and $\tau_{\varepsilon,M}^x := \tau_\varepsilon^x \wedge \tau_M^x$ for all $0 < \varepsilon < x < M$. By Itô's formula,

$$s\left(X_{t\wedge\tau_{\varepsilon,M}^x}^x\right) = s(x) + \int_0^{t\wedge\tau_{\varepsilon,M}^x} As(X_u)\,du$$

$$+ \int_0^{t\wedge\tau_{\varepsilon,M}^x} s'(X_u)\sigma\sqrt{X_u}\,dB_u$$

$$= s(x) + \int_0^{t\wedge\tau_{\varepsilon,M}^x} s'(X_u)\sigma\sqrt{X_u}\,dB_u. \qquad [3.29]$$

The derivative s' is bounded in the interval $[\varepsilon, M]$; moreover, it is bounded from below by a positive constant. The same holds for the values of the process X_t till the moment $\tau_{\varepsilon,M}^x$. Therefore, there exist constants $0 < c < C$ (depending on ε and M) such that the integrand process $Y_u = s'(X_u)\sigma\sqrt{X_u}$ satisfies the inequalities

$$c \leqslant Y_u \leqslant C, \quad 0 \leqslant u \leqslant \tau_{\varepsilon,M}^x.$$

Therefore, $Y \in \mathcal{H}^2[0, t \wedge \tau_{\varepsilon,M}^x]$ for all $0 < \varepsilon < x < M$ and $t \geqslant 0$ and, thus, the expectation of the integral in equation [3.29] equals zero. Therefore, we have

$$\mathbf{E}s\left(X_{t\wedge\tau_{\varepsilon,M}^x}^x\right) = s(x), \quad 0 < \varepsilon < x < M. \qquad [3.30]$$

Moreover,

$$\mathrm{Var}\,s\left(X_{t\wedge\tau_{\varepsilon,M}^x}^x\right) = \mathbf{E}\left(\int_0^{t\wedge\tau_{\varepsilon,M}^x} Y_u\,dB_u\right)^2$$

$$= \mathbf{E}\int_0^{t\wedge\tau_{\varepsilon,M}^x} Y_u^2\,du \geqslant c^2\mathbf{E}\int_0^{t\wedge\tau_{\varepsilon,M}^x} du$$

$$= c^2\mathbf{E}\left(t \wedge \tau_{\varepsilon,M}^x\right), \quad 0 < \varepsilon < x < M.$$

As $\varepsilon \leqslant X^x_{t \wedge \tau^x_{\varepsilon,M}} \leqslant M$ and the function s is bounded in the interval $[\varepsilon, M]$, the right-hand side of the inequality is bounded by some constant $\widetilde{C}$ independent of t. Therefore, we have

$$\mathbf{E}s\left(t \wedge \tau^x_{\varepsilon,M}\right) \leqslant \frac{\widetilde{C}}{c^2}, \quad t \geqslant 0.$$

Letting $t \to \infty$, we get $\mathbf{E}(\tau^x_{\varepsilon,M}) \leqslant \frac{\widetilde{C}}{c^2} < \infty$, and therefore, $\tau^x_{\varepsilon,M} < \infty$ a.s. (for all $0 < \varepsilon < x < M$). Moreover, since $s(\varepsilon) \leqslant s\left(X_{t \wedge \tau^x_{\varepsilon,M}}\right) \leqslant s(M)$, we can apply the Lebesgue theorem passing to the limit, as $t \to \infty$, in equation [3.30]:

$$\mathbf{E}s\left(X^x_{\tau^x_{\varepsilon,M}}\right) = s(x), \quad 0 < \varepsilon < x < M. \tag{3.31}$$

As the random variable $X^x_{\tau^x_{\varepsilon,M}}$ may take only two values, ε (when $\tau^x_\varepsilon < \tau^x_M$) and M (when $\tau^x_M < \tau^x_\varepsilon$), from the last equality, we get

$$s(x) = s(\varepsilon)\mathbf{P}\{\tau^x_\varepsilon < \tau^x_M\} + s(M)\mathbf{P}\{\tau^x_M < \tau^x_\varepsilon\}. \tag{3.32}$$

Denote $s(0) := \lim_{x \downarrow 0} s(x)$ and consider the cases in the formulation of the theorem.

(1) $a \geqslant \sigma^2/2$. Then $2a/\sigma^2 \geqslant 1$, and therefore, $s(0) := -\int_0^1 y^{-2a/\sigma^2} e^{\frac{2b}{\sigma^2}y}\,\mathrm{d}y = -\infty$. Hence,

$$\mathbf{P}\{\tau^x_0 < \tau^x_M\} = \lim_{\varepsilon \downarrow 0} \mathbf{P}\{\tau^x_\varepsilon < \tau^x_M\}$$

$$= \lim_{\varepsilon \downarrow 0} \frac{s(x) - s(M)\mathbf{P}\{\tau^x_M < \tau^x_\varepsilon\}}{s(\varepsilon)} = 0,$$

where the first equality follows from the continuity of probability and the equality for events

$$\{\tau^x_0 < \tau^x_M\} = \bigcup_{\varepsilon > 0}\{\tau^x_\varepsilon < \tau^x_M\} = \lim_{\varepsilon \downarrow 0}\{\tau^x_\varepsilon < \tau^x_M\}.$$

Now, since

$$\{\tau^x_0 < \infty\} = \bigcup_{M > 0}\{\tau^x_0 < \tau^x_M\} = \lim_{M \to \infty}\{\tau^x_0 < \tau^x_M\},$$

letting $M \to \infty$, we get

$$\mathbf{P}\{\tau_0^x < \infty\} = \lim_{M \to \infty} \mathbf{P}\{\tau_0^x < \tau_M^x\} = 0.$$

(2,3) $0 \leqslant a < \sigma^2/2$. Then, $2a/\sigma^2 < 1$ and $s(0) := -\int_0^1 y^{-2a/\sigma^2} e^{\frac{2b}{\sigma^2} y} \, dy > -\infty$, and from equation [3.32], we get

$$s(x) = s(0)\mathbf{P}\{\tau_0^x < \tau_M^x\} + s(M)\mathbf{P}\{\tau_M^x < \tau_0^x\}, \quad 0 < x < M,$$

and thus

$$\mathbf{P}\{\tau_M^x < \tau_0^x\} = \frac{s(x) - s(0)\mathbf{P}\{\tau_0^x < \tau_M^x\}}{s(M)}, \quad 0 < x < M.$$

Therefore, since

$$\{\tau_0^x = \infty\} = \bigcap_{M > 0} \{\tau_M^x < \tau_0^x\} = \lim_{M \to \infty} \{\tau_M^x < \tau_0^x\},$$

it follows that

$$\mathbf{P}\{\tau_0^x = \infty\} = \lim_{M \to \infty} \mathbf{P}\{\tau_M^x < \tau_0^x\}$$

$$= \lim_{M \to \infty} \frac{s(x) - s(0)\mathbf{P}\{\tau_0^x < \tau_M^x\}}{s(M)}.$$

If $b \geqslant 0$ (case 2), then $s(\infty) := \lim_{M \to \infty} s(M) = \infty$, and from the last equality, we get $\mathbf{P}\{\tau_0^x = \infty\} = 0$.

If $b < 0$ (case 3), then $s(\infty) < \infty$, and, consequently,

$$\mathbf{P}\{\tau_0^x = \infty\} = \lim_{M \to \infty} \frac{s(x) - s(0)\mathbf{P}\{\tau_0^x < \tau_M^x\}}{s(M)}$$

$$= \frac{s(x) - s(0)\mathbf{P}\{\tau_0^x < \infty\}}{s(\infty)}$$

$$= \frac{s(x) - s(0)(1 - \mathbf{P}\{\tau_0^x = \infty\})}{s(\infty)},$$

that is,

$$\mathbf{P}\{\tau_0^x = \infty\} = \frac{s(x) - s(0)}{s(\infty) - s(0)} \in (0, 1). \qquad \Box$$

3.3.1. *Moments and covariance of the CIR process*

Although the distributions of the values of the CIR process are known, their expressions are rather complicated. Therefore, the CIR process is usually modeled by its approximations. However, it is interesting that it is rather easy to calculate the moments of this process. For example, taking the expectation in equation [3.28] written in the integral form,

$$X_t = x + \int_0^t (a - bX_s)\, ds + \sigma \int_0^t \sqrt{X_s}\, dB_s, \qquad [3.33]$$

we get

$$m_1(t, x) := \mathbf{E}X_t = x + \int_0^t (a - b\,\mathbf{E}X_s)\, ds$$

$$= x + \int_0^t \big(a - b\,m_1(s, x)\big)\, ds$$

or (after differentiation with respect to t)

$$m_1'(x, t) = a - b\,m_1(t, x), \quad m_1(0, x) = x.$$

The solution of this equation is

$$m_1(t, x) = xe^{-bt} + \frac{a}{b}(1 - e^{-bt}),\ b \neq 0;\ m_1(t, x) = x + at,\ b = 0.$$

Note that for $b > 0$, $m_1(t, x) \to a/b$ as $t \to \infty$, or, equivalently, for the process [3.27], $\mathbf{E}r_t \to b$ as $t \to \infty$. This property of the CIR process is called mean reversion.

The higher-order moments can be obtained by applying a recurrent formula. Using Itô's formula, we have

$$X_t^n = x^n + n \int_0^t X_s^{n-1} dX_s + \frac{n(n-1)}{2} \int_0^t X_s^{n-2} d\langle X \rangle_s$$

$$= x^n + n \int_0^t X_s^{n-1}(a - bX_s)\, ds + n \int_0^t X_s^{n-1}\sigma\sqrt{X_s}\, dB_s$$

$$+ \frac{n(n-1)}{2} \int_0^t X_s^{n-2}\sigma^2 X_s\, ds$$

$$= x^n + \int_0^t (an + \frac{n(n-1)}{2}\sigma^2) X_s^{n-1}\, ds - bn \int_0^t X_s^n\, ds$$

$$+ n\sigma \int_0^t X_s^{n-1/2}\, dB_s.$$

Taking the expectations, we have

$$m_n(t, x) := \mathbf{E} X_t^n = x^n + \int_0^t \left(an + \frac{n(n-1)}{2}\sigma^2\right) m_{n-1}(s, x)\, ds$$

$$- bn \int_0^t m_n(s, x)\, ds.$$

By differentiating with respect to t and solving the obtained differential equation with respect to m_n, we get the recurrent relation

$$m_n(t, x) = e^{-nbt}\left(x^n + \int_0^t \left(an + \frac{n(n-1)}{2}\sigma^2\right)\right.$$

$$\left. e^{nbs} m_{n-1}(s, x)\, ds\right).$$ [3.34]

For example, for $b \neq 0$, we have

$$\mathbf{E} X_t^2 = m_2(t, x) = e^{-2bt}\left(x^2 + \int_0^t (2a + \sigma^2) e^{2bs} m_1(s, x)\, ds\right)$$

$$= e^{-2bt}\left(x^2 + \int_0^t (2a + \sigma^2) e^{2bs}\left(xe^{-bs} + \frac{a}{b}(1 - e^{-bs})\right) ds\right)$$

$$= e^{-2bt}\left(x^2 + (2a + \sigma^2)\int_0^t \left((x - \frac{a}{b})e^{bs} + \frac{a}{b}e^{2bs}\right) ds\right)$$

$$= e^{-2bt}\left(x^2 + (2a + \sigma^2)\left((x - \frac{a}{b})(e^{bt} - 1)/b\right.\right.$$

$$\left.\left. + \frac{a}{2b^2}(e^{2bt} - 1)\right)\right)$$

$$= x^2 e^{-2bt} + \frac{2a + \sigma^2}{2b^2}(2(bx - a)(e^{-bt} - e^{-2bt})$$

$$+ a(1 - e^{-2bt}))$$

$$= x^2 e^{-2bt} + x\frac{2a + \sigma^2}{b}(e^{-bt} - e^{-2bt})$$

$$+ a\frac{2a + \sigma^2}{2b^2}(1 - e^{-bt})^2$$

and

$$\operatorname{Var} X_t = \mathbf{E}X_t^2 - (\mathbf{E}X_t)^2 = x\frac{\sigma^2}{b}(e^{-bt} - e^{-2bt}) + \frac{a\sigma^2}{2b^2}(1 - e^{-bt})^2.$$

REMARK 3.3.– Note that, in the limit, $\mathbf{E}X_t \to \frac{a}{b}$, $\mathbf{E}X_t^2 \to \frac{a\sigma^2}{2b^2} + \frac{a^2}{b^2}$, and $\operatorname{Var} X_t \to \frac{a\sigma^2}{2b^2}$ as $t \to \infty$. In fact, we can say even more: as $t \to \infty$, the distribution of X_t tends to the so-called gamma distribution with density

$$p(y) = \frac{\beta^\alpha}{\Gamma(\alpha)}y^{\alpha-1}\exp\{-\beta y\}, \quad y > 0,$$

with scale and rate parameters $\alpha = 2a/\sigma^2$ and $\beta = 2b/\sigma^2$. This means that the density $p(t, x, y)$ of X_t (starting at $X_0 = x$) tends to $p(y)$ for all $y > 0$ as $t \to \infty$.

Now, let us calculate the covariance function

$$C(s, t) = \operatorname{Cov}(X_s, X_t) = \mathbf{E}\big[(X_s - \mathbf{E}X_s)(X_t - \mathbf{E}X_t)\big], \quad s, t \geqslant 0.$$

First, denote $\tilde{X}_t := X_t - a/b$. Note that the processes X and $\tilde{X}$ differ by a constant and, therefore, have the same covariance function. The process $\tilde{X}$ satisfies the equation

$$d\tilde{X}_t = -b\tilde{X}_t\,dt + \sigma\sqrt{\tilde{X}_t + \frac{a}{b}}\,dB_t. \quad \tilde{X}_0 = \tilde{x} = x - \frac{a}{b}. \qquad [3.35]$$

Applying the integration-by-parts formula (theorem 1.18) for $\tilde{X}_t e^{bt}$, we have

$$\tilde{X}_t e^{bt} = \tilde{x} + \int_0^t \tilde{X}_u b e^{bu}\,du + \int_0^t e^{bu}\Big(-b\tilde{X}_u\,du$$

$$+ \sigma\sqrt{\tilde{X}_u + \frac{a}{b}}\,dB_u\Big)$$

$$= \tilde{x} + \sigma\int_0^t e^{bu}\sqrt{\tilde{X}_u + \frac{a}{b}}\,dB_u.$$

and thus

$$\tilde{X}_t = \tilde{x}e^{-bt} + \sigma e^{-bt}\int_0^t e^{bu}\sqrt{\tilde{X}_u + \frac{a}{b}}\,dB_u.$$

Therefore, for $0 \leqslant s \leqslant t$,

$$
\begin{aligned}
C(s,t) &= \sigma^2 e^{-b(s+t)} \mathbf{E}\bigg(\int_0^s e^{bu} \sqrt{\tilde{X}_u + \frac{a}{b}} \, \mathrm{d}B_u \\
&\quad \cdot \int_0^t e^{bu} \sqrt{\tilde{X}_u + \frac{a}{b}} \, \mathrm{d}B_u \bigg) \\
&= \sigma^2 e^{-b(s+t)} \mathbf{E} \int_0^s e^{2bu} \Big(\tilde{X}_u + \frac{a}{b} \Big) \, \mathrm{d}u \qquad \text{[Theorem 1.6 (4)]}
\end{aligned}
$$

$$
\begin{aligned}
&= \sigma^2 e^{-b(s+t)} \int_0^s e^{2bu} \Big(\mathbf{E}\tilde{X}_u + \frac{a}{b} \Big) \, \mathrm{d}u \\
&= \sigma^2 e^{-b(s+t)} \int_0^s e^{2bu} \Big(\tilde{x} e^{-bu} + \frac{a}{b} \Big) \, \mathrm{d}u \\
&= \sigma^2 e^{-b(s+t)} \int_0^s \Big(\tilde{x} e^{bu} + \frac{a}{b} e^{2bu} \Big) \, \mathrm{d}u \\
&= \sigma^2 e^{-b(s+t)} \Big(\frac{\tilde{x}}{b} (e^{bs} - 1) + \frac{a}{2b^2} (e^{2bs} - 1) \Big) \\
&= \sigma^2 \Big(\frac{\tilde{x}}{b} (e^{-bt} - e^{-b(s+t)}) + \frac{a}{2b^2} (e^{-b(t-s)} - e^{-b(s+t)}) \Big).
\end{aligned}
$$

Note that, in the limit, the covariance depends on the difference $t - s$ only. Namely,

$$
C(s, s+t) \to C(t) := \frac{a\sigma^2}{2b^2} e^{-t}, \qquad s \to \infty.
$$

3.3.2. *The value of zero-coupon bond in the CIR model*

Now, let us come back to equation [3.27]. Assuming that the risk premium is of the form $\lambda_t = \bar{\lambda}\sqrt{r_t}$, the spot rates with respect to risk-neutral probability satisfy the equation of the same form,

$$
\mathrm{d}r_t = \tilde{a}(\tilde{b} - r_t) \, \mathrm{d}t + \sigma\sqrt{r_t} \, \mathrm{d}\tilde{B}_t, \qquad\qquad [3.36]
$$

where $\tilde{a} = a + \sigma\bar{\lambda}$ and $\tilde{b} = ab/(a + \sigma\bar{\lambda})$.

THEOREM 3.3.– The value of zero-coupon bond in the CIR model is

$$P(t, T) = f(T - t) \exp \{-g(T - t)r_t\},$$

where the functions f and g are defined by

$$g(\theta) = \frac{2(e^{\rho\theta} - 1)}{(\tilde{a} + \rho)(e^{\rho\theta} - 1) + 2\rho}, \quad \rho = \sqrt{\tilde{a}^2 + 2\sigma^2};$$

$$f(\theta) = \phi(\theta)^{2\tilde{a}\tilde{b}/\sigma^2}, \quad \phi(\theta) = \frac{2\rho e^{(\tilde{a}+\rho)\theta/2}}{(\tilde{a} + \rho)(e^{\rho\theta} - 1) + 2\rho}.$$

The yield curve at moment t is

$$R(t, T) = \frac{g(T - t)}{T - t} r_t - \frac{2\tilde{a}\tilde{b}}{\sigma^2} \frac{\ln \phi(T - t)}{T - t}.$$

PROOF.– The PDE [3.12] for $P(t, T) = p(t, r_t, T)$ in the CIR model is

$$\frac{\partial p}{\partial t} + \tilde{a}(\tilde{b} - r) \frac{\partial p}{\partial r} + \frac{\sigma^2 r}{2} \frac{\partial^2 p}{\partial r^2} = r p \qquad [3.37]$$

with end condition $p(T, r, T) = 1$. Let us look for a solution in the form $p(t, r, T) = f(\theta)e^{-g(\theta)r}$, $\theta = T - t$. Putting it into the PDE and dividing by $e^{-g(\theta)r}$, we get

$$- f'(\theta) + f(\theta)g'(\theta)r - \tilde{a}(\tilde{b} - r)f(\theta)g(\theta) + \frac{\sigma^2 r}{2} f(\theta)g^2(\theta)$$

$$- rf(\theta) = 0, f(\theta) \left(g'(\theta) + \tilde{a}g(\theta) + \frac{\sigma^2}{2}g^2(\theta) - 1 \right) r$$

$$- \left(f'(\theta) + \tilde{a}\tilde{b}f(\theta)g(\theta) \right) = 0.$$

Equating the terms at the first- and zero-order powers of r differential equations, we get the differential equation system

$$g'(\theta) + \tilde{a}g(\theta) + \frac{\sigma^2}{2}g^2(\theta) - 1 = 0,$$

$$f'(\theta) + \tilde{a}\tilde{b}f(\theta)g(\theta) = 0$$

with initial conditions $g(0) = 0$ and $f(0) = 1$. We solve it in a standard (and rather tedious) way. For the first equation, we have

$$-\int_0^{g(\theta)} \frac{\mathrm{d}x}{\frac{\sigma^2}{2}x^2 + \tilde{a}x - 1} = \theta.$$

Applying the formula

$$\int \frac{\mathrm{d}x}{ax^2 + bx + c} = \frac{1}{\sqrt{D}} \ln \left| \frac{2ax + b - \sqrt{D}}{2ax + b + \sqrt{D}} \right| \quad (D := b^2 - 4ac > 0),$$

we obtain for $g = g(\theta)$:

$$\frac{1}{\rho} \ln \frac{\sigma^2 x + \tilde{a} + \rho}{-(\sigma^2 x + \tilde{a} - \rho)} \bigg|_{x=0}^{g} = \theta, \quad \rho := \sqrt{\tilde{a}^2 + 4\frac{\sigma^2}{2}}$$

$$= \sqrt{\tilde{a}^2 + 2\sigma^2} > \tilde{a};$$

$$\ln \frac{\sigma^2 g + \tilde{a} + \rho}{-(\sigma^2 g + \tilde{a} - \rho)} - \ln \frac{\tilde{a} + \rho}{-(\tilde{a} - \rho)} = \rho\theta;$$

$$\frac{\sigma^2 g + \tilde{a} + \rho}{\rho - \tilde{a} - \sigma^2 g} = \frac{\tilde{a} + \rho}{\rho - \tilde{a}} e^{\rho\theta};$$

$$\frac{\sigma^2 g + \tilde{a}}{\rho} = \frac{(\tilde{a} + \rho)e^{\rho\theta} + (\tilde{a} - \rho)}{(\tilde{a} + \rho)(e^{\rho\theta} - 1) + 2\rho};$$

$$g = \frac{1}{\sigma^2} \left[\rho \frac{(\tilde{a} + \rho)e^{\rho\theta} + (\tilde{a} - \rho)}{(\tilde{a} + \rho)(e^{\rho\theta} - 1) + 2\rho} - \tilde{a} \right]$$

$$= \frac{1}{\sigma^2} \frac{\rho(\tilde{a} + \rho)e^{\rho\theta} + \rho(\tilde{a} - \rho) - \tilde{a}(\tilde{a} + \rho)(e^{\rho\theta} - 1) - 2\rho\tilde{a}}{(\tilde{a} + \rho)(e^{\rho\theta} - 1) + 2\rho}$$

$$= \frac{1}{\sigma^2} \frac{\rho(\tilde{a} + \rho)(e^{\rho\theta} - 1) - \tilde{a}(\tilde{a} + \rho)(e^{\rho\theta} - 1)}{(\tilde{a} + \rho)(e^{\rho\theta} - 1) + 2\rho}$$

$$= \frac{1}{\sigma^2} \frac{(\rho^2 - \tilde{a}^2)(e^{\rho\theta} - 1)}{(\tilde{a} + \rho)(e^{\rho\theta} - 1) + 2\rho}$$

$$= \frac{2(e^{\rho\theta} - 1)}{(\tilde{a} + \rho)(e^{\rho\theta} - 1) + 2\rho}.$$

The function f is found by integrating the obtained expression of g:

$$\ln f(\theta) = -\tilde{a}\tilde{b} \int_0^\theta g(\tau)\, d\tau = -\tilde{a}\tilde{b} \int_0^\theta \frac{2(e^{\rho\tau} - 1)}{(\tilde{a} + \rho)(e^{\rho\tau} - 1) + 2\rho}\, d\tau$$

$$= -\tilde{a}\tilde{b} \cdot \frac{2}{\tilde{a} + \rho} \int_0^\theta \frac{e^{\rho\tau} - 1}{e^{\rho\tau} + \frac{\rho - \tilde{a}}{\rho + \tilde{a}}}\, d\tau.$$

Applying the formula

$$\int \frac{e^{a\tau} + c}{e^{a\tau} + b}\, d\tau = \frac{c}{b}\tau + \frac{b - c}{ab} \ln(e^{a\tau} + b),$$

we continue:

$$\ln f(\theta) = -\frac{2\tilde{a}\tilde{b}}{\tilde{a} + \rho}\left(-\frac{\rho + \tilde{a}}{\rho - \tilde{a}}\theta + \frac{2}{\rho - \tilde{a}}\left[\ln\left(e^{\rho\theta} + \frac{\rho - \tilde{a}}{\rho + \tilde{a}} \right) \right.\right.$$

$$\left.\left. - \ln\left(1 + \frac{\rho - \tilde{a}}{\rho + \tilde{a}} \right) \right] \right)$$

$$= \frac{\tilde{a}\tilde{b}(\rho + \tilde{a})}{\sigma^2}\theta - \frac{2\tilde{a}\tilde{b}}{\sigma^2} \ln\left(\frac{(\rho + \tilde{a})(e^{\rho\theta} - 1) + 2\rho}{2\rho} \right).$$

From this, we finally get

$$f(\theta) = \exp\left\{ \frac{\tilde{a}\tilde{b}(\rho + \tilde{a})}{\sigma^2}\theta \right\} \left(\frac{2\rho}{(\rho + \tilde{a})(e^{\rho\theta} - 1) + 2\rho} \right)^{2\tilde{a}\tilde{b}/\sigma^2}$$

$$= \left[\frac{2\rho e^{(\rho + \tilde{a})\theta/2}}{(\tilde{a} + \rho)(e^{\rho\theta} - 1) + 2\rho} \right]^{2\tilde{a}\tilde{b}/\sigma^2}.$$

We derive the yield curve by using, as usual, the equality $R(t, T) = -\frac{1}{T-t} \ln P(t, T)$. Figures 3.2 and 3.3 show typical graphs of yield curves for different parameter sets.

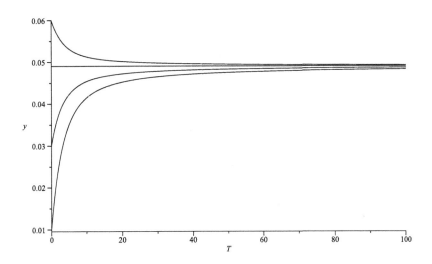

Figure 3.2. *Yield curves $R(0, T)$ in the CIR model, $T \in [0; 100]$:*
$\sigma = 0.1; \tilde{b} = 0.05; \tilde{a} = 0.5; R_\infty \approx 0.049; r_0 = 0.01; \ 0.03; \ 0.06$

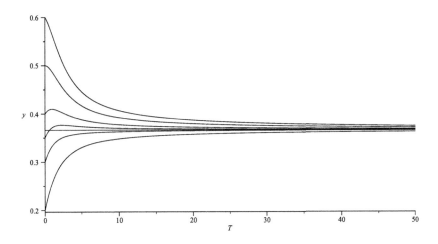

Figure 3.3. *Yield curves $R(0, T)$ in the CIR model, $T \in [0; 50]$: $\sigma = 0.5$;*
$\tilde{b} = 0.5; \tilde{a} = 0.5; R_\infty \approx 0.366; r_0 = 0.2; \ 0.3; \ 0.4; \ 0.5; \ 0.6$

Alternative method. We have

$$P(t, T) = \widetilde{\mathbf{E}} \left[\exp \left\{ - \int_t^T r_s \, ds \right\} \bigg| \mathcal{F}_t \right] = \widetilde{\mathbf{E}} \left[\exp \left\{ - \int_t^T r_s \, ds \right\} \bigg| r_t \right]$$

$$= F(T - t, r_t)$$

with $F(\theta, x) = \tilde{\mathbb{E}} \exp\{-\int_0^\theta r_u^x \, du\}$, where r^x is the solution of equation [3.36] with initial value $r_0 = x$. By the Feynman–Kac formula (corollary 1.3(2)), F satisfies the equation

$$\frac{\partial F(\theta, x)}{\partial \theta} = AF(\theta, x) - xF(\theta, x)$$

$$= \tilde{a}(\tilde{b} - x)\frac{\partial F(\theta, x)}{\partial x} + \frac{\sigma^2 x}{2}\frac{\partial^2 F(\theta, x)}{\partial x^2} - xF(\theta, x)$$

with initial condition $F(0, x) = 1$. Note that this is the same, already solved, equation [3.37] for the bond value $p(t, r, T) = F(T - t, r)$. Thus, we obtain the same result.

For more about the CIR process, see [COX 85], section 6.2.2.

EXERCISE 3.11.– Solve the CIR equation

$$dX_t = \left(\frac{\sigma^2}{2} - aX_t\right) dt + \sigma\sqrt{X_t}\, dB_t, \quad X_0 = x > 0.$$

Hint: Denoting $Y_t = \sqrt{X_t}$ and using Itô's formula, reduce the equation to the Vašíček model.

EXERCISE 3.12.– Let $\tilde{r}_t$ and $\hat{r}_t$ be two independent CIR processes with parameters $(a, \tilde{b}, \sigma)$ and $(a, \hat{b}, \sigma)$, respectively (see equation [3.27]). Show that $r_t := \tilde{r}_t + \hat{r}_t$ is also a CIR process with parameters $(a, \tilde{b} + \hat{b}, \sigma)$.

REMARK 3.1.– You will need the Lévy theorem on characterization of the Brownian motion (theorem 1.22).

EXERCISE 3.13.– Let r_t^i, $i = 1, \ldots, d$, be solutions of the equations

$$dr_t^i = -ar_t^i \, dt + \sigma \, dB_t^i$$

with independent Brownian motions B^i, $i = 1, \ldots, d$. Denote

$$\tilde{r}_t := |r_t|^2 = \sqrt{\sum_{i=1}^d (r_t^i)^2}.$$

Show that $\tilde{r}_t$ is a CIR process that solves the equation

$$d\tilde{r}_t = 2a\left(\frac{d\sigma^2}{2a} - \tilde{r}_t\right) dt + 2\sigma\sqrt{\tilde{r}_t}\, d\tilde{B}_t$$

with a one-dimensional Brownian motion $\widetilde{B}$.

REMARK 3.2.– Again, you will need the Lévy theorem.

EXERCISE 3.14.– Affine interest rate models are described by equations of the form

$$dr_t = \tilde{\alpha}(t, r_t)\, dt + \sigma(t, r_t)\, d\widetilde{B}_t$$

with coefficients $\tilde{\alpha}(t, r) = \alpha_1(t)\, r + \alpha_2(t)$ and $\sigma(t, r) = \sqrt{\sigma_1^2(t)\, r + \sigma_2^2(t)}$, where $\widetilde{B}$ is a Brownian motion with respect to risk-neutral probability, and α_1, α_2. σ_1, σ_2 are non-random functions.

Show that the zero-coupon price in such a model is

$$P(t, T) = p(t, r_t, T), \quad 0 \leqslant t \leqslant T,$$

where

$$p(t, r, T) = \exp\{c(t)r + d(t)\}$$

with coefficients $c = c(t)$ and $d = d(t)$ satisfying the differential equations

$$\begin{cases} c'(t) + \alpha_1(t)c(t) + \frac{\sigma_1^2(t)}{2}c^2(t) - 1 = 0, \\ d'(t) + \alpha_2(t)c(t) + \frac{\sigma_2^2(t)}{2}c^2(t) = 0, \end{cases}$$

and end conditions $c(T) = d(T) = 0$.

3.4. The Heath–Jarrow–Morton model

In the previous models, the yield curves depend only on one variable, the spot rate r_t, and the value of zero-coupon bond satisfies equation [3.14] (p. 89), i.e.

$$dP(t, T) = P(t, T)\left[r_t\, dt + \gamma(t, r_t, T)\, d\widetilde{B}_t\right].$$

This approach is rather restrictive. In the Heath–Jarrow–Morton model, it is supposed that, instead of the dynamics of the spot rate r_t and the volatility of the form $\gamma(t, r_t, T)$, we are given the volatility as a general known random process.

We consider a no-arbitrage market with a zero-coupon bond and riskless bond S_t^0 with spot rate r_t. We suppose that, under the risk-neutral probability $\widetilde{\mathbf{P}}$, the discounted value of the zero-coupon bond $P(t, T)/S_t^0$ is a martingale and that the value of zero-coupon bond $P(t, T)$ satisfies the equation

$$\frac{\mathrm{d}P(t, T)}{P(t, T)} = r_t \, \mathrm{d}t + \sigma(t, T) \, \mathrm{d}\widetilde{B}_t, \quad P(T, T) = 1, \qquad [3.38]$$

where $\widetilde{B}$ is a Brownian motion with respect to $\widetilde{\mathbf{P}}$, and $\sigma(t, T)$, $t \leqslant T$, is the volatility function, which is a family of adapted random processes called local volatilities, depending on parameter T. Owing to the end condition $P(T, T) = 1$, we suppose that $\sigma(T, T) = 0$. We also assume that σ is continuous and continuously differentiable with respect to T. Then, clearly, its derivative

$$\Sigma(t, T) = \frac{\partial \sigma(t, T)}{\partial T}, \; t \leqslant T,$$

is also an adapted process. A specific property of the HJM model is that its yield curve (with respect to risk-neutral probability) is completely determined by the volatility function.

THEOREM 3.4.– i) In the HJM model, the value of zero-coupon bond is

$$P(t, T) = P_f(0, t, T) \exp\left\{ \int_0^t \left(\sigma(s, T) - \sigma(s, t) \right) \mathrm{d}\widetilde{B}_s \right.$$
$$\left. - \frac{1}{2} \int_0^t \left(\sigma^2(s, T) - \sigma^2(s, t) \right) \mathrm{d}s \right\}, \qquad [3.39]$$

where $P_f(0, t, T) = P(0, T)/P(0, t)$ is the forward price of the bond at moment t.

ii) The yield curve at moment t is

$$R(t, T) = R_f(0, t, T) + \frac{1}{2} \int_0^t \frac{\sigma^2(s, T) - \sigma^2(s, t)}{T - t} \mathrm{d}s$$
$$- \int_0^t \frac{\sigma(s, T) - \sigma(s, t)}{T - t} \mathrm{d}\widetilde{B}_s,$$

where $R_f(0,t,T) = -\frac{1}{T-t}\ln P_f(0,t,T) = -\frac{1}{T-t}\ln\frac{P(0,T)}{P(0,t)}$ is the average forward price of the bond at moment t.

iii) The forward spot rate at moment t is

$$f(t,T) = f(0,T) + \int_0^t \sigma(s,T)\Sigma(s,T)\,ds - \int_0^t \Sigma(s,T)\,d\tilde{B}_s; \quad [3.40]$$

in particular, the spot rate is

$$r_t = f(t,t) = f(0,t) + \int_0^t \sigma(s,t)\Sigma(s,t)\,ds - \int_0^t \Sigma(s,t)\,d\tilde{B}_s. \quad [3.41]$$

PROOF.–

i) Denote $Y_t = \int_0^t r_s\,ds + \int_0^t \sigma(s,T)\,d\tilde{B}_s$. Then, from equation [3.38] and example 1.2 , we get

$$P(t,T) = P(0,T)\exp\left\{Y_t - \frac{1}{2}\langle Y\rangle_t\right\}$$

$$= P(0,T)\exp\left\{\int_0^t r_s\,ds + \int_0^t \sigma(s,T)\,d\tilde{B}_s - \frac{1}{2}\int_0^t \sigma^2(s,T)\,ds\right\}.$$

From this, taking $T = t$, we have

$$1 = P(0,t)\exp\left\{\int_0^t r_s\,ds + \int_0^t \sigma(s,t)\,d\tilde{B}_s - \frac{1}{2}\int_0^t \sigma^2(s,t)\,ds\right\}.$$

By dividing the equalities obtained with each other, we eliminate the process r and obtain expression [3.39].

ii) The yield curve is obtained directly by substituting into the equality $R(t,T) = -\frac{1}{T-t}\ln P(t,T)$ the expression of $P(t,T)$ obtained in part (i).

iii) $f(t,T) = -\dfrac{\partial \ln P(t,T)}{\partial T} = -\dfrac{\partial \ln P_f(0,t,T)}{\partial T}$

$$-\dfrac{\partial\left(\int_0^t (\sigma(s,T)-\sigma(s,t))\,d\tilde{B}_s - \frac{1}{2}\int_0^t (\sigma^2(s,T)-\sigma^2(s,t))\,ds\right)}{\partial T}$$

$$= f(0,T) - \int_0^t \frac{\partial \sigma(s,T)}{\partial T} \, \mathrm{d}\widetilde{B}_s + \frac{1}{2} \int_0^t \frac{\partial \sigma^2(s,T)}{\partial T} \, \mathrm{d}s$$

$$= f(0,T) - \int_0^t \Sigma(s,T) \, \mathrm{d}\widetilde{B}_s + \int_0^t \sigma(s,T)\Sigma(s,T) \, \mathrm{d}s. \quad [3.42]$$

$\square$

As an example, let us consider the case $\sigma(s,t) = \sigma(t-s)$ with a constant σ. Then, by equation [3.40], we have

$$f(t,T) = f(0,T) + \sigma^2 \int_0^t (T-s) \, \mathrm{d}s - \sigma \widetilde{B}_t$$

$$= f(0,T) + \sigma^2 t \left(T - \frac{t}{2} \right) - \sigma \widetilde{B}_t$$

and, in particular,

$$r_t = f(t,t) = f(0,t) + \frac{\sigma^2 t^2}{2} - \sigma \widetilde{B}_t. \quad [3.43]$$

We see that the spot rate may (unfortunately) take negative values with a positive probability. By equation [3.2], we also have

$$P(t,T) = \exp\left(-\int_t^T f(t,s) \, \mathrm{d}s \right)$$

$$= \exp\left(-\int_t^T f(0,s) \, \mathrm{d}s - \sigma^2 \int_t^T t\left(s - \frac{t}{2} \right) \mathrm{d}s + \sigma(T-t)\widetilde{B}_t \right)$$

$$= \exp\left(-\int_t^T f(0,s) \, \mathrm{d}s - \sigma^2 \frac{Tt}{2}(T-t) + \sigma(T-t)\widetilde{B}_t \right).$$

$$[3.44]$$

A shortcoming of this formula is that, in reality, $\widetilde{B}_t$ is not observable. The situation can be corrected by multiplying equation [3.43] by $T - t$:

$$\sigma(T-t)\widetilde{B}_t = (T-t)f(0,t) + \frac{\sigma^2 t^2(T-t)}{2} - (T-t)r_t.$$

Putting the last expression into equation [3.44], we get

$$P(t,T) = \exp\left(-\int_t^T \left(f(0,s) - f(0,t)\right) \mathrm{d}s\right.$$
$$\left. -\frac{\sigma^2 t(T-t)^2}{2} - (T-t)r_t\right).$$

In this formula, there "participate" only observable random variables, and we need to estimate only one unknown parameter σ.

EXERCISE 3.15.– Justify the differentiation with respect to T under the sign of stochastic integral in equation [3.42].

EXERCISE 3.16.– The HJM model is often defined in terms of the forward spot rate, which is supposed to satisfy an SDE of the form

$$\mathrm{d}f(t,T) = \alpha(t,T)\,\mathrm{d}t + \Sigma(t,T)\,\mathrm{d}\widetilde{B}_t.$$

Prove that, in the no-arbitrage market, $\alpha(t,T)$ must satisfy the relation

$$\alpha(t,T) = \Sigma(t,T)\int_t^T \Sigma(t,s)\,\mathrm{d}s, \quad 0 \leqslant t \leqslant T.$$

(Compare with equation [3.40].)

> Don't cry because it came to an end, smile because it happened.
> *Gabriel Garcia Marquez*

Bibliography

Certain authors, speaking of their works, say "My book", "My commentary", "My history", etc. They resemble middle-class people who have a house of their own, and always have "My house" on their tongue. They would do better to say, "Our book", "Our commentary", "Our history", etc., because there is in them usually more of other people's than their own.

Blaise Pascal, Thoughts

[BLA 73] BLACK F., SCHOLES M., "The pricing of options and corporate liabilities", *Journal of Political Economy*, vol. 81, pp. 635–654, 1973.

[BOR 02] BORODIN A.N., SALMINEN P., *Handbook of Brownian Motion – Facts and Formulae*, 2nd ed., Springer, 2002.

[COX 85] COX J.C., INGERSOLL J.E., ROSS S.A., "A theory of the term structure of interest rates", *Econometrica*, vol. 53, pp. 385–407, 1985.

[DAN 07] DANA R.-A., JEANBLANC M., *Financial Markets in Continuous Time*, Springer, 2007.

[DEL 06] DELBAEN F., SCHACHERMAYER W., *The Mathematics of Arbitrage*, Springer Finance, 2006.

[LAM 96] LAMBERTON D., LAPEYERE B., *Introduction to Stochastic Calculus Applied to Finance*, Chapman & Hall, 1996.

[MAC 11] MACKEVIČIUS V., *Introduction to Stochastic Analysis: Integrals and Differential Equations*, ISTE Ltd, London and John Wiley & Sons, New York, 2011.

[MIK 99] MIKOSCH T., *Elementary Stochastic Calculus with Finance in View*, World Scientific Publishing, Singapore, 1999.

[PHA 07] PHAM H., Introduction aux Mathématiques et Modèles Stochastiques des Marchés Financieres, Lecture Notes, University of Paris 7, Version 2006–2007.

[PRI 12] PRIVAULT N., *An Elementary Introduction to Stochastic Interest Rate Modeling*, World Scientific Publishing, Singapore, 2012.

[STE 01] STEELE J.M., *Stochastic Calculus and Financial Applications*, Springer, 2001.

[WIL 06] WILLIAMS R.J., *Introduction to the Mathematics of Finance*, AMS, Providence, 2006.

Index

Printed in the United States
By Bookmasters